Milestones in Drug Therapy
MDT

Series Editors

Prof. Dr. Michael J. Parnham
PLIVA
Research Institute
Prilaz baruna Filipovica 25
10000 Zagreb
Croatia

Prof. Dr. J. Bruinvels
INFARM
Sweelincklaan 75
NL-3723 JC Bilthoven
The Netherlands

Antidepressants

Edited by B.E. Leonard

Springer Basel AG

Editor

Prof. Brian E. Leonard
Pharmacology Department
University College Galway
Galway
Ireland

Advisory Board

J.C. Buckingham (Imperial College School of Medicine, London, UK)
D. de Wied (Rudolf Magnus Institute for Neurosciences, Utrecht, The Netherlands)
F.K. Goodwin (Center on Neuroscience, Washington, USA)
G. Lambrecht (J.W. Goethe Universität, Frankfurt, Germany)

Library of Congress Cataloging-in-Publication Data

Antidepressants / edited by B.E. Leonard.
 p. ; cm. – (Milestones in drug therapy)
 Includes bibliographical references and index.
 ISBN 978-3-0348-9526-2   ISBN 978-3-0348-8344-3 (eBook)
 DOI 10.1007/978-3-0348-8344-3
   1. Antidepressants. 2. Depression, Mental. I. Leonard, B, E. II. Series.
   [DNLM: 1. Antidepressive Agents – therapeutic use. 2. Depressive Disorder – drug
 therapy. 3. Antidepressive Agents – pharmacology. 4. Depressive
 Disorder – physiopathology. QV 77.5 A6288 2000]
 RM332, A5732 2000
 616.85'27061–dc21   00-057948

Deutsche Bibliothek Cataloging-in-Publication Data

Antidepressants / ed. by B.E. Leonard. – Basel ; Boston ; Berlin : Birkhäuser, 2001
 (Milestones in drug therapy)
 ISBN 978-3-0348-9526-2

The publisher and editor can give no guarantee for the information on drug dosage and administration contained in this publication. The respective user must check its accuracy by consulting other sources of reference in each individual case.
The use of registered names, trademarks etc. in this publication, even if not identified as such, does not imply that they are exempt from the relevant protective laws and regulations or free for general use.
This work is subject to copyright. All rights are reserved, whether the whole or part of the material is concerned, specifically the rights of translation, reprinting, re-use of illustrations, recitation, broadcasting, reproduction on microfilms or in other ways, and storage in data banks. For any kind of use, permission of the copyright owner must be obtained.

© 2001 Birkhauser Verlag AG
Originally published by Birkhaüser Verlag, Basel - Boston - Berlin in 2001
Softcover reprint ofthe hardcover 1st edition 2001
Printed on acid-free paper produced from chlorine-free pulp. TCF ∞
Cover illustration: Flesinoxan HCl (see chapter Andrews/Pinder)

ISBN 978-3-0348-9526-2

9 8 7 6 5 4 3 2 1

Contents

List of contributors VII

Brian E. Leonard
Preface IX

Clinical issues

Kerstin Bingefors, Lisa Ekselius and Lars von Knorring
Long-term course, comorbidity and long-term outcome
in depressive disorders 3

Jules Angst and Hans H. Stassen
Do antidepressants really take several weeks to show effect? 21

Stephen M. Stahl
Commentary on the limitation of antidepressants in current use . . 31

Mechanisms of action

Caroline McGrath and Trevor R. Norman
Mechanisms underlying the speed of onset
of antidepressant response 47

Ian A. Paul
Calcium signaling cascades, antidepressants
and major depressive disorders 63

Markers of depression

Timothy G. Dinan
The hypothalamic-pituitary-adrenal axis
and antidepressant action 83

Philip J. Cowen
Neuroendocrine markers of depression
and antidepressant drug action 95

Brian E. Leonard
Brain cytokines and the psychopathology of depression 109

The future of antidepressants

John S. Andrews and Roger M. Pinder
Chemistry and pharmacology of novel antidepressants 123

Subject index 147

List of contributors

John Stuart Andrews, Janssen Research Foundation, Turnhoutseweg 30, B-2340 Beerse, Belgium; e-mail: JANDREW1@janbe.jnj.com

Jules Angst, Psychiatric University Hospital Zurich, Research Department, P. O. Box 68, 8029 Zurich, Switzerland; e-mail: k454910@bli.unizh.ch

Kerstin Bingefors, Dept. of Neuroscience, Psychiatry, Uppsala University, University Hospital, 75185 Uppsala, Sweden;
e-mail: Chris.Bingefors@samfarm.uu.se

Philip J. Cowen, University department of psychiatry, Warneford Hospital, Oxford 0X3 7JX, UK; e-mail: phil.cowen@psych.ox.ac.uk

Timothy G. Dinan, Department of Psychiatry, Royal College of Surgeons in Ireland, St. Stephen's Green, Dublin 2, Ireland; e-mail: tdinan@indigo.ie

Lisa Ekselius, Dept. of Neuroscience, Psychiatry, Uppsala University, University Hospital, 75185 Uppsala, Sweden;
e-mail: Lisa.Ekselius@UASPsyk.uu.se

Caroline McGrath, Dept. Psychiatry, University of Melbourne, Australia;
e-mail: mcgrath@clyde.its.unimelb.edu.au

Brian E. Leonard, Pharmacology department, National University of Ireland, Galway, Ireland; e-mail: belucg@iol.ie

Trevor R. Norman, Dept. Psychiatry, University of Melbourne, Australia;
e-mail: trevor@austin.unimelb.edu.au

Ian A. Paul, Department of Psychiatry, University of Mississippi Medical Center, 2500 North State St., Jackson, MS 39216-4505, USA;
e-mail: ipaul@psychiatry.umsmed.edu

Roger M. Pinder, Organon Inc., 375 Mount Pleasant Avenue, West Orange, New Jersey 07059, USA; e-mail: r.pinder@organoninc.com

Stephen M. Stahl, Clinical Neuroscience Research Center, 8899 University Center Lane, San Diego, California 92122;
e-mail: smstahl1@compuserve.com

Hans H. Stassen, Psychiatric University Hospital Zurich, Research Department, P. O. Box 68, 8029 Zurich, Switzerland;
e-mail: k454910@bli.unizh.ch

Lars von Knorring, Dept. of Neuroscience, Psychiatry, Uppsala University, University Hospital, 75185 Uppsala, Sweden;
e-mail: Lars.von_Knorring@UASPsyk.uu.se

Preface

Some three hundred years ago, Robert Burton described in his famous thesis *The Anatomy of Melancholy* depression as a tapestry of symptoms which differ from person to person and which are never constant within the same individual. And yet, inspite of the advances in our understanding of the psychobiology of depression in the past thirty years, and the increasing number of biologically oriented psychiatrists being trained in our medical schools, the impact of the advances in the management of depressed patients has been at best modest. Depression, in its complexity of presentation, continues to confound many clinicians and probably represents one of the world's greater epidemics. Indeed, it has been estimated by the World Health Organisation that the affective disorders will be one of the top causes of premature death and ill health in the world by 2025, surpassing most infectious and physical diseases that are gradually being defeated by new drug treatments combined with improvements in diet and in reduced environmental pollution.

Despite the increasing availability of effective antidepressants, depression continues to be one of the most misunderstood, unrecognised and untreated illnesses worldwide. It has been estimated that the incidence may be as high as 20% of the population, with about 5% of patients experiencing a major depressive episode necessitating appropriate antidepressant therapy. Even when recognised, the depressed patient is frequently given inadequate doses of the antidepressant and thereby experiences side effects with little therapeutic benefit. This results in a growing population of untreated patients, proportion of whom are often labelled "treatment resistant". An untold number of depressed patients thereby go untreated and are condemned unnecessarily to live miserable lives exposed to ill health due to autoimmune diseases, cancers, heart disease and infections. In addition, depression can be a fatal disease when leading to suicide. It has been estimated that about 70% of the patients who commit suicide in industrialized countries have visited a physician up to the month before their demise. Perhaps this is not surprising when it is realized that in the USA alone fewer than 20% of depressed patients visiting a psychiatrist are prescribed the appropriate antidepressant treatment. The situation for the depressed patient visiting a general practitioner must be considerably worse.

One of the reasons for this situation relates to the paucity of training in all areas of psychopharmacology at both the undergraduate and postgraduate levels. Apart from the provision of appropriate courses in psychopharmacology, textbooks and monographs providing up-to-date information on the advances being made in the diagnosis and treatment of depression can help to stimulate further research and thereby improve treatment.

With these thoughts in mind, I enlisted the help of eleven leading research groups in both preclinical and clinical research to present succinct overviews of the advances in their fields of interest. The first section of the monograph is devoted to important clinical issues, namely the long-term outcome of treating depression (Kerstin Bingefors and colleagues), the time of onset of the antidepressant response (Jules Angst and H. Strassen) and the limitation of antidepressants in current use (Stephen Stahl). Fundamental aspects of the mechanisms of action of antidepressants form the second section of the monograph. In this section, Ian Paul covers the role of second messengers in the antidepressant response, while Caroline McGrath and Trevor Norman discuss the mechanisms that affect the speed of onset of antidepressants. The third section considers the role of the neuroendocrine axis in the antidepressant response (by Ted Dinan) and how the changes in the hormonal status can act as markers of treatment response (Phil Cowen). This section concludes by my own chapter on the role of proinflammatory cytokines in the psychobiology of depression. The final section is an important chapter by John Andrews and Roger Pinder on the antidepressants of the future, a chapter in which the authors critically assess the novel antidepressants that are currently in development.

Hopefully, this monograph will help not only to educate but also to stimulate research the biology of depression and of antidepressant's mode of action. Only in this way it will be possible to advance our knowledge to enable a rapid and accurate diagnosis of depression and hopefully to develop novel treatments that are more effective than those currently available.

B.E. Leonard, Galway

Clinical issues

Long-term course, comorbidity and long-term outcome in depressive disorders

Kerstin Bingefors, Lisa Ekselius, Lars von Knorring

The Uppsala group for the Study of Affective Disorders (USAD), Dept. of Neuroscience, Psychiatry, Uppsala University, University Hospital, SE-751 85 Uppsala, Sweden

Introduction

In old textbooks the knowledge about depressive disorders was often based upon in-patient samples. In later years much of the knowledge has been based on results from controlled randomised trials in which new antidepressants have been tested. However, there are many differences between patients included in clinical trials and patients in clinical practice [1]. In clinical practice psychiatric comorbidity is common, while those patients often are excluded from clinical trials. In the same way, somatic comorbidity is common in clinical practice [2], while those patients often are excluded in clinical trials.

Depressed patients included in epidemiological studies or studied in natural settings have a high psychiatric comorbidity [2], somatic comorbidity [2], comorbidity with personality disorders [3], and an increased mortality [4, 5] even if suicide deaths are disregarded.

The aim of the present paper is to try to summarise our present knowledge about psychiatric and somatic comorbidity in depressive disorders, to evaluate the long-term outcome and mortality, and to discuss possible mechanisms for the reported increased mortality.

Comorbidity, axis I – psychiatric disorders

In epidemiological studies or studied in natural settings depressed patients tend to have a higher mean utilisation of psychiatric health care per year and higher use of psychotropic drugs than non-depressed subjects in the general population [2], indicating a higher general psychiatric morbidity and a higher psychiatric comorbidity (Tab. 1).

In a Spanish study [6] the aim was to estimate the current prevalence of DSM-III-R [7] and ICD-10 [8] psychiatric disorders in 18-year-old members of the general population. Subjects were assessed using the Schedules for Clinical Assessment in Neuropsychiatry (SCAN). Nearly 30% of the study subjects reported at least one current disorder according to ICD-10

Table 1. Mean psychiatric health care utilisation per year, hospital care and primary care among depressed and non-depressed subjects in the general population [2] and percent of patients in the general population with and without depression, respectively, receiving prescription drugs by drug class [2]

	Controls 25–44 years	Depressives 25–44 years	p	Controls 45–64 years	Depressives 45–64 years	p	Controls 65–74 years	Depressives 65–74 years	p
Men									
Hospital care – psychiatric bed days	0.2	0.7	***	0.7	2.7	***	1.6	3.9	***
Women									
Hospital care – psychiatric bed days	0.0	3.5	***	0.6	2.2	***	1.0	3.8	
Men									
CNS	10.3	100.0	***	17.1	100.0	***	25.5	100.0	***
– psychotropics	3.8	48.6	***	7.5	75.0	***	15.9	82.3	***
– neuroleptics	0.7	14.3	***	1.3	37.5	***	1.7	26.5	***
– tranquilizers	2.3	31.4	***	4.0	50.0	***	7.7	53.0	***
– sedatives	1.5	8.6	*	3.6	39.6	***	9.4	44.1	***
Women									
CNS	14.6	100.0	***	25.5	100.0	***	35.5	100.0	***
– psychotropics	5.9	57.0	***	15.9	80.7	***	27.0	75.9	***
– neuroleptics	0.7	15.2	***	1.9	24.8	***	2.8	18.4	***
– tranquilizers	4.2	26.6	***	9.9	53.2	***	17.4	57.5	***
– sedatives	1.8	26.6	***	7.8	39.4	***	13.2	34.5	***

criteria, and almost 21% reported at least one current disorder according to DSM-III-R criteria. Women had a significantly higher probability of suffering from a psychiatric disorder than men. The most common disorders were insomnia, dysthymia, major depression and simple phobia. Nearly 40% of the diagnosed subjects had one or more comorbid disorders. Comorbidity was found to be higher among female subjects. Consistent with previous risk factor research, it was found that women had higher rates of mood, anxiety and sleep disorders than men.

The overlap of symptoms associated with comorbid anxiety and depressive disorders makes diagnosis, research, and treatment particularly difficult [9]. Recent evidence suggests genetic and neurobiologic similarities between depressive and anxiety disorders. Comorbid depression and anxiety are highly prevalent conditions. Patients with panic disorder, generalised anxiety disorder, social phobia, and other anxiety disorders are also frequently clinically depressed. Approximately 85% of patients with depression also experience significant symptoms of anxiety. Similarly, comorbid depression occurs in up to 90% of patients with anxiety disorders. Patients with comorbid disorders do not respond as well to therapy, have a more protracted course of illness, and experience less positive treatment outcomes. One key to successful treatment of patients with mixed depressive and anxiety disorders is early recognition of comorbid conditions. Antidepressant medications, including the selective serotonin reuptake inhibitors, tricyclic antidepressants, and monoamine oxidase inhibitors, are effective in the management of comorbid depression and anxiety. The high rates of comorbid depression and anxiety argue for well-designed treatment studies in these populations.

Thus, anxiety and depressive symptoms commonly co-occur but the underlying mechanisms for this covariation remain poorly understood. Genetic strategies are a useful means of investigating whether the comorbidity of two sets of symptoms or disorders can be explained by the same etiological factors. In one study [10] a systematically ascertained sample of 172 twin pairs aged 8 to 16 years was examined regarding the causes of covariation of maternally rated anxiety and depressive symptoms. The results suggest that most of the covariation can be explained by a common set of genes that influence anxiety and depressive symptoms. Some covariation between anxiety and depressive symptoms is also explained by environmental influences of the non-shared type. In addition, depressive symptoms also appear to be influenced by specific genetic factors.

In a study by Rodney et al. [11] that explored the effect of comorbid anxiety disorders in patients admitted to an inpatient specialist Mood Disorders Unit for the treatment of a primary major depressive episode, subjects were assessed on admission and discharge. DSM-III-R diagnoses for major depression and anxiety disorders were established using CIDI-Auto; comorbid anxiety disorders were coexistent in time with the major depression, with both conditions meeting diagnostic criteria at the time of assessment. Severity of illness was assessed using the Hamilton Depression/Melancholia Scale, the revised Hamilton Anxiety Scale and the revised Beck Depression Inventory. The study cohort was divided into three groups: depression alone (n = 33), one comorbid anxiety disorder (n = 15), and two or more comorbid anxiety disorders (n = 24). No particular anxiety disorder predominated. Interestingly the presence or absence of comorbid anxiety with severe major depression made no significant difference to treatment choice or outcome results. Specifically there was no significant difference between the three groups in the utilisation of electroconvulsive therapy and pharmacotherapy (including antidepressants, benzodiazepines and neuroleptics); all subjects improved significantly on both depression and anxiety ratings, and length of inpatient stay did not vary significantly between the three groups. Thus, the existence of comorbid anxiety disorders in those patients who presented for treatment of a primary major depressive episode did not significantly effect choice of treatment or treatment outcome, suggesting that there is a close interrelationship between the two conditions [11].

In a study [12] in which 650 depressed outpatients visiting general medical clinicians and mental health specialists were followed for 1 or 2 years the effect of a comorbid anxiety disorder was evaluated. All types of anxiety increased the probability of a new depressive episode among patients with subthreshold depression. Co-occurring panic and phobia decreased the likelihood of remission. The initial number of depressive symptoms was greatest among depressed patients with comorbid anxiety and this relatively higher level persisted over 2 years. The findings emphasise the poor clinical prognosis associated with comorbid anxiety disorder.

Thus, international epidemiological and clinical studies have shown that comorbid depression and anxiety is of major importance, resulting in more severe symptoms, impairment, subjective distress, and longitudinal course than either anxiety or depression alone. Current evidence demonstrates the importance of evaluating both threshold and subthreshold levels of depression and anxiety [13].

There is also a high comorbidity between many common psychiatric disorders and substance abuse disorders. In the National Comorbidity Survey [14], 5877 respondents were asked about the history of five psychiatric disorders in their parents: major depression (MD), generalised anxiety disorder (GAD), antisocial personality disorder (ASP), alcohol abuse/dependence (AAD) and drug abuse/dependence (DAD). Significant familial aggregation was seen for all disorders. Controlling for other disorders produced only modest reductions in the odds ratios for MD, GAD and AAD and larger reductions for ASP and DAD. The familial transmission of these disorders can be explained by underlying vulnerabilities to internalising and to externalising disorders transmitted across generations with moderate fidelity. Thus, familial aggregation of common psychiatric and substance use disorders is substantial in epidemiological samples. The examined environmental adversities account for little of the observed parent-offspring transmission of these conditions.

While the extent of medical comorbidity in depressive illness is widely recognised, there have been few studies where objective and subjective assessment of comorbidity in chronic depression have been made. In a group of outpatients with chronic depressive symptoms [15] the extent of subjective and objective comorbidity was assessed. Eighty-seven outpatients with a history of depressed mood for at least 2 years were assessed using the SCID and the Hamilton Depression scale. Physician and patient assessment of any medical comorbidity were completed. Sixty percent of patients, and significantly more public general hospital outpatients than private outpatients, viewed themselves as suffering from a serious medical illness in addition to their depression. Only 8% were viewed objectively as having a serious medical condition. The discrepancy between patient and physician assessments of medical comorbidity in chronic depression is of note and may relate to depressed mood.

Comorbidity, axis II – personality disorders

The association of major depressive disorders with personality disorders (Axis II) is relevant in terms of clinical, therapeutic and prognostic aspects. The DSM-IV [16] defines personality disorders in terms of habitual and enduring patterns of perception, cognition and behaviour that are relatively inflexible and maladaptive, and lead to significant functional impairment or subjective distress. The disorder must manifest by late adolescence or early adulthood and not be the consequence of another mental disorder, chronic intoxication, or a general medical condition. The DSM-IV personality disorders are subdivided into three descriptive clusters. Cluster A includes the disorders that share some phenomenologic similarity with schizophrenia: paranoid, schizoid and schizotypal personality disorders. Antisocial, borderline, histrionic and narcissistic personality disorders are grouped within Cluster B characterised by emotional instability and dramatic or impulsive, erratic behaviour. Cluster C consists of avoidant, dependent, and obsessive compulsive personality disorders, which have features in common with anxiety disorders. In addition, a tentative Axis II diagnosis of depressive personality disorder is included in the DSM-IV appendix.

Comorbidity between depression and personality disorders is widespread and extremely strong. Studies reporting the prevalence of personality disorders in depressed patients generally suggest that 20% to 50% of inpatients and 50% to 85% of out-patients with a current major depressive disorder have an associated personality disorder [17]. In a series of 400 major depressed patients treated by general practitioners recruited for a clinical trial of two SSRIs the presence of a personality disorder was assessed using the SCID screen questionnaire [18]. A comorbid personality disorder was found in 68% of the male and in 60% of the female patients suggesting a high comorbidity also in primary care patients (Tab. 2).

The relationship between personality disorder and depression may stem from at least three different processes [19, 20]: 1) Characterological predisposition refers to the tendency for people with certain maladaptive personality traits, such as neuroticism or excessive interpersonal dependency, to be at greater risk for development of depression. 2) Complication describes the development or exaggeration of personality traits as a product of a single depressive episode or several recurring episodes. 3) Attenuation; this model presumes that personality disorders are an attenuated or alternative expression of the disease process which underlies the depressive disorder. Both personality and depressive disorders are seen as arising from the same genetic or constitutional origins. Empirical evidence can be found in support of each of the processes described here. Thus, the association between personality and depression is complex and multifactorial.

With respect to the relationship between depression, comorbid personality disorders, and response to treatment, the majority of studies find that the

Table 2. Comorbid personality disorders in a series of 400 patients with major depressive disorders treated by general practitioners recruited for a clinical trial of two SSRIs. The presence of a personality disorder determined by means of the SCID screen questionnaire

	Males (n = 112)	Females (n = 288)	
Paranoid PD	35 (31.3%)	78 (27.1%)	$\chi^2 = $ 0.69, n.s.
Schizotypal PD	6 (5.4%)	5 (1.7%)	$\chi^2 = $ 3.95, $p < 0.05$
Schizoid PD	4 (3.6%)	4 (1.4%)	$\chi^2 = $ 1.96, n.s.
Cluster A PD	39 (34.8%)	82 (28.5%)	$\chi^2 = $ 1.54, n.s.
Histrionic PD	8 (7.1%)	17 (5.9%)	$\chi^2 = $ 0.21, n.s.
Narcissistic PD	28 (25.0%)	29 (10.1%)	$\chi^2 = $ 14.71, $p < 0.001$
Borderline PD	21 (18.8%)	61 (21.2%)	$\chi^2 = $ 0.29, n.s.
Cluster B PD	38 (33.9%)	81 (28.1%)	$\chi^2 = $ 1.30, n.s.
Avoidant PD	48 (42.9%)	103 (35.8%)	$\chi^2 = $ 1.73, n.s.
Dependent PD	27 (24.1%)	50 (17.4%)	$\chi^2 = $ 2.36, n.s.
Obsessive-Compulsive PD	28 (25.0%)	69 (24.0%)	$\chi^2 = $ 0.05, n.s.
Passive-Aggressive PD	25 (22.3%)	22 (7.6%)	$\chi^2 = $ 16.77, $p < 0.001$
Cluster C PD	63 (56.3%)	145 (50.4%)	$\chi^2 = $ 1.13, n.s.
Any PD	76 (67.9%)	173 (60.1%)	$\chi^2 = $ 2.08, n.s.

presence of personality disorders is related to both poor response to treatment [21–24] and a poorer prognosis for long-term outcome [25, 26]. However, there are some studies which show that the presence of a personality disorder does not impair clinical outcome to specific treatments [27–29]. In a later study in primary care [30] depressed patients with a personality disorder responded worse to 16 weeks treatment than patients with no personality disorder. However, after 18 months there were no differences in ratings of depression between the groups. From that study it was concluded that the presence of a personality disorder delays recovery from major depressive illness.

A number of suicide studies that have investigated personality disorders show varying rates of such morbidity, from 3% [31] to 57% [32]. The relationships between personality disorders and suicide were recently investigated among two aboriginal groups and the Han Chinese in East Taiwan [33]. Biographical reconstructive interviews were conducted for consecutive suicides from each of the three ethnic groups. In all groups, a high proportion of those who commited suicide suffered from personality disorder before suicide (47–78%), and the most prevalent category was the ICD-10 emotionally unstable personality disorder [8]. The risk for suicide was mainly significantly associated with emotionally unstable personality disorder, comorbidity among personality disorders, and comorbidity of personality disorder with other psychiatric disorders, particularly severe depression.

Comorbidity, axis III – somatic disorders

Affective disorders are common in patients with somatic disease. In epidemiological studies, depressive syndromes have been demonstrated in 59% of primary care attendees and in 22–33% of inpatients in general medical wards. A study by Silverstone [34] found substantially lower rates of major depressive disorder (5.1%) in inpatients when using the modified criteria suggested by Endicott [35], which replace the four somatic symptoms in DSM-III-R with nonsomatic symptoms. However, as a high comorbidity between depressive disorders and somatic disease has been found also when antidepressant treatment, long-term outcome or course have been considered, it might be that the modified criteria suggested by Endicott are too narrow.

The problems of underdetection and inadequate treatment is substantial in patients with concurrent somatic illness. It has been demonstrated that only 14–38% of somatically ill patients receive a correct diagnosis of depression and that even fewer are prescribed antidepressants. There has been considerable discussion about whether the high comorbidity of depression and somatic disease is a consequence of the somatic condition, or whether the depressive symptoms precede the medical diagnosis. Several plausible explanations for depression and somatic conditions occurring together were summarised by the Depression Guideline Panel [36].

- The somatic condition biologically causes the depression;
- The somatic disease precipitates the onset of depression in those genetically prone to depression;
- The somatic disease psychologically causes the depression;
- The somatic disease and the mood disorder are not causally related.

Perris [37] put forward a concept of general vulnerability as an explanation for the high somatic morbidity found in depressed patients.

A number of studies have shown that individuals suffering from depression or depressive symptoms are high utilisers of health care and that they are as functionally impaired as patients with chronic somatic disorders. A population based study by Bingefors et al. [2] showed that antidepressant-treated depressed patients in the community used a disproportionately high amount of health care for somatic conditions (Tab. 3), compared to the population in the community studied. Patients with depression generally used more than twice as much health care for somatic complaints than the general population. Furthermore, they also used considerably more non-psychotropic prescription drugs (Fig. 1). Patients treated for depression received between 2–10 times as many non-antidepressant prescriptions per person compared to other individuals in the population. The high use of prescription drugs was not explained by use of one or a few particular drug groups. Polypharmacy was common; the risk for receiving drugs from five or more of the main pharmacological classes was 3.7 (C.I. (95%) 2.3–5.9)

Table 3. Mean non-psychiatric health care utilisation per year, hospital care and primary care among depressed and non-depressed subjects in the general population [2]

	Controls 25–44 years	Depressives 25–44 years	Sign.	Controls 45–64 years	Depressives 45–64 years	Sign.	Controls 65–74 years	Depressives 65–74 years	Sign.
Men									
Hospital care – total bed days	0.5	1.4	***	1.9	5.9	***	5.4	19.5	***
Ambulatory care visits	1.5	4.9	***	2.4	6.6	***	3.3	5.0	***
GP visits	1.0	2.3	***	1.2	3.2	***	1.7	3.2	***
– without psychiatry	1.0	1.7	**	1.2	2.6	***	1.6	2.6	**
No. prescriptions	1.5	11.0	***	3.2	19.0	***	6.8	17.3	***
– without antidepressants	1.5	6.9	***	3.2	15.3	***	6.8	14.0	***
– without psychotropics	1.4	4.3	**	2.9	9.4	***	6.2	9.7	**
Women									
Hospital care – total bed days	1.0	4.9	**	1.7	4.5	***	4.6	10.5	
Ambulatory care visits	2.4	6.6	***	2.7	6.0	***	3.3	5.2	***
GP visits	1.5	3.0	***	1.6	3.3	***	2.0	3.0	***
– without psychiatry	1.4	2.4	**	1.5	2.4	***	2.0	2.3	
No. prescriptions	2.7	13.5	***	4.7	20.4	***	8.0	20.1	***
– without antidepressants	2.7	10.1	***	4.7	16.4	***	8.0	16.4	***
– without psychotropics	2.6	7.1	**	4.2	9.3	***	7.2	11.6	**

for depressed men and 3.1 (C.I. (95%) 2.4–3.9) for depressed women. The higher use of health care and drugs usually found in females was not seen in this study; depressed men had as much excess use as women. The high use of prescription drugs among depressed patients is not only an indicator of somatic comorbidity, it may also be a serious cause for concern in itself. Polypharmacy leads to a number of complications, such as increased problems with adverse reactions, interactions and non-compliance leading to direct and indirect health care costs. Studies suggest that the treatment response to antidepressants in patients with somatic comorbidity is less than that in patients with depression only [38, 39].

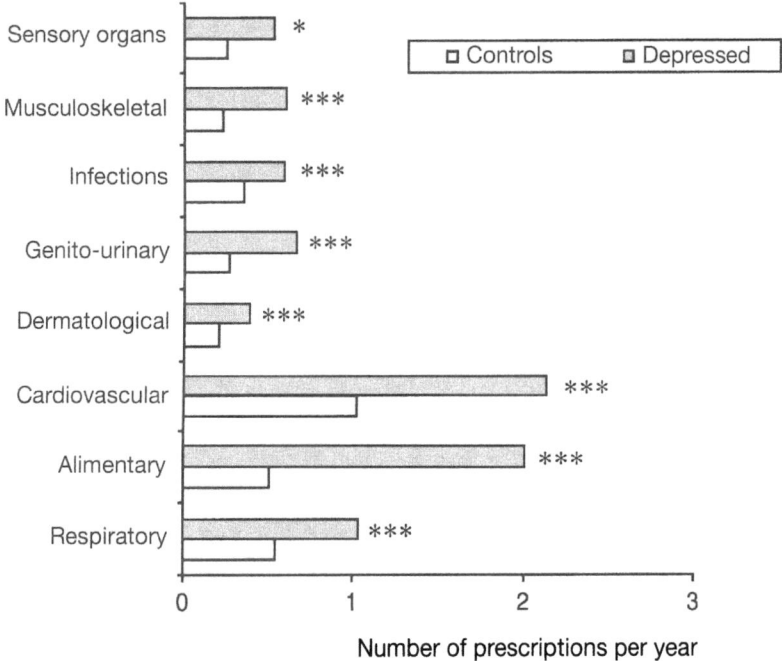

Figure 1. Prescription of drugs in different anatomical therapeutic classification groups prescribed during a year to depressed patients and controls in the community [2]. Analysis of variance, $*p < 0.05$, $**p < 0.01$, $***p < 0.001$.

Long-term course

There have been few longitudinal studies on health care utilisation among depressed patients, particularly use of health services prior to a diagnosis of depression. Widmer and Cadoret [40] found an increase in the number of physician visits and hospitalisation prior to a developing depression, while a British study of a general practitioner's patients found an excess of consultations for physical illness for a considerable period before a diagnosis of depression was made. The excess of consultations continued after treatment and recognition of the psychiatric condition [41].

A study by Bingefors et al. [42] followed prescription drug use in all patients in a community treated for the first time with antidepressants for 5 years before and 6 years after the index treatment. Compared to a population control group, patients with depression had a higher average use of somatic drugs during all 11 years of the study (Fig. 2). Five years before the index date the depressed patients used on average 7.2 somatic prescriptions as compared to 4.1 for referents. Use of somatic drugs among the depressed increased significantly ($p < 0.05$) during the first year after index date and then fell back to the same level as before treatment. The high use of drugs was not limited to only one particular drug or even a few classes of

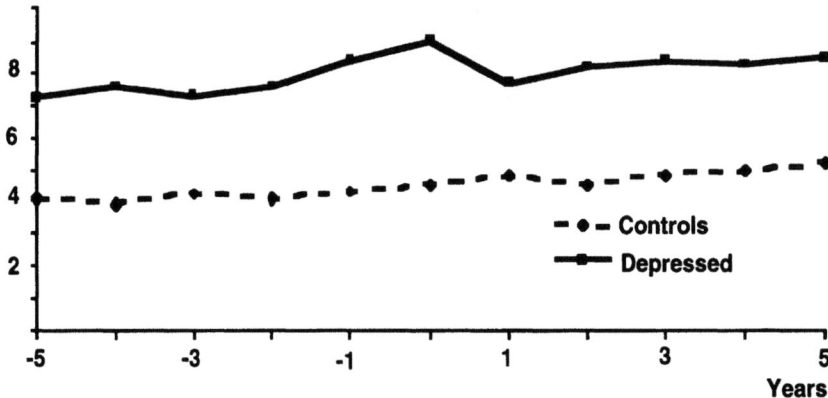

Figure 2. Average use of non-psychotropic drugs during 5 years before and 6 years after the index episode (first episode with antidepressant treatment) [42].

drugs. There was no difference between genders in excess drug use, indicating a similar comorbidity in man and women.

In another study by Bingefors et al. [2], use of primary health care was studied during the same time period as the prescription drug use. The depressed patient group had a significantly higher rate of primary care visits than the referent group (Fig. 3) during all of the 11 years of the study. There was a significant ($p < 0.05$) increase in the year immediately prior to first treatment of depression. The two diagnostic groups with the highest ratio of excess visits were digestive diseases (ICD-IX) and symptoms and ill- defined conditions (ICD-XIV).

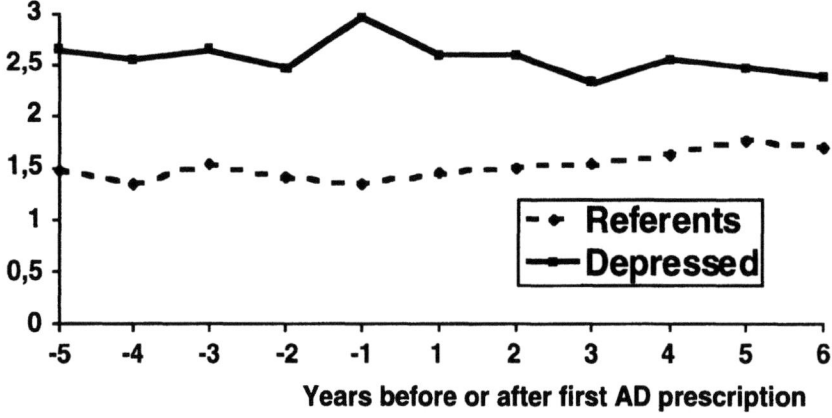

Figure 3. Number of visits to general practitioners due to somatic disorders during 5 years before and 6 years after the index episode (first episode with antidepressant treatment) [65].

Long-term outcome

Mortality is the most commonly used measure of health status and outcome in populations. Consequently, mortality in depressed patients has been of long-term interest to researchers. Studies on different patient groups, with or without concurrent physical health problems, have included mortality directly related to depression, such as suicides and accidental deaths, as well as from natural causes [43]. In particular, depression has been implicated as a risk factor in cardiovascular disease [44, 45]. In a study of bipolar (manic-depressive and recurrent depressive psychoses) [4] Perris was able to demonstrate that both male and female, and both unipolar and bipolar patients had a shortening of the expected lifetime of 4–23%. The mean expectancy of life patient/population was for bipolar males 0.82, bipolar females 0.77, unipolar males 0.91 and unipolar females 0.96. In summary, the remaining mean expectation of life being 26% lower in bipolar and 4% lower in unipolar patients than in the general population. Among the causes of death, only suicide (22%) differed from the general population. It was suggested that the higher mortality rate in bipolar patients, which seemed to be independent of suicide or other causes of death, may reflect a lower general resistance of the organism [37].

Most studies have followed hospitalised patients or patients treated in psychiatric care facilities while there have been relatively few studies of mortality in connection with depression in the community where 80% of patients are treated. An American study [46] reported that depression increased mortality during a 9-year period in a community sample aged 40 and older. Bingefors et al. [5] studied the mortality in the earlier mentioned population-based cohort of first time depressed patients. Nine-year mortality was significantly higher ($p < 0.01$) in the depressed group as compared to population controls (Fig. 4). The increased mortality was due to an increase among patients 65 years and older. Mortality was increased in both sexes. There was no significant difference in the distribution of deaths on main causes of death. However, suicides were more common among the depressed patients, but that did not explain the overall excess in mortality. In a multivariate analysis mortality was related to baseline somatic health conditions at date of first depression treatment (Fig. 4). Depression remained a risk factor for mortality even after controlling for these baseline conditions. Depression as a risk factor was on the same magnitude as that of some of the most common chronic diseases (Fig. 5).

In a recent review article, Harris and Barraclough [47] have summarised the excess mortality for the different psychiatric disorders. As concerns affective disorders, the Standardised Mortality Ratio (SMR) is significantly increased (SMR = 171*). As expected the highest SMR is for suicide (SMR = 1990*). However, also natural causes of death had a clearly significant increase in SMR (134*). The causes of death were infectious

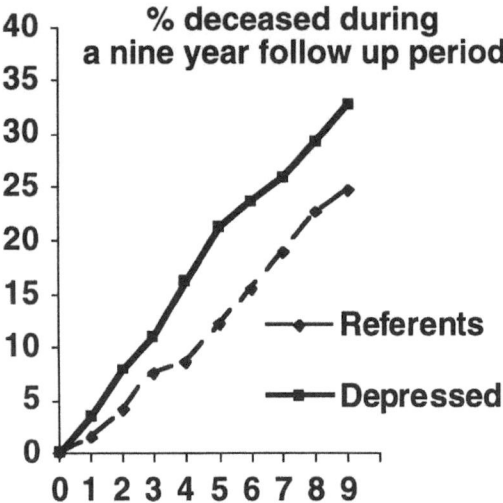

Figure 4. Mortality during a 9-year follow-up period of depressed patients treated with antidepressants [5].

(SMR = 131*), circulatory (SMR = 116*) and respiratory (SMR = 194*). The analysis is based on 25 reported studies on a total population of 18 997 from nine countries with follow-up periods ranging from 0–50 years. Fifty-three percent of the population were male and Sweden accounted for 28% of the all cause expected value.

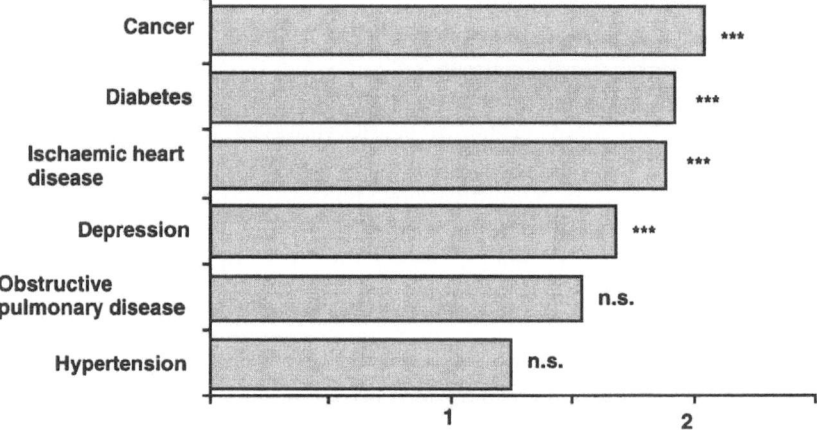

Figure 5. Adjusted mortality rate in a Cox regression analysis relating age, sex and baseline morbidity to 9-years mortality in depressed outpatients treated with antidepressants [5].

Possibilities to decrease the long-term morbidity and mortality

In an editorial in Lancet [48] in 1992, it was discussed if depression and suicide are preventable or not. It was concluded that there is a substantial case for chronic treatment of mood disorders as endorsed by a WHO consensus statement. However, it was also regarded of interest that all reports of successful long-term treatment results at the time came from special mood disorders clinics. For example, it was demonstrated by Coppen et al. [49] that at a mood disorders clinic where most patients were treated with lithium and where compliance was good, the standardised mortality ratio SMR was decreased to 0.6 in an 11 year follow-up. In the same way, it was demonstrated in a four-nation study that patients at special lithium clinics had a SMR of 0.9 [50]. However, patients not treated at the lithium clinics but at the same hospital still had increased mortality at follow-up [48]. Thus, knowledge among the caregivers seems to be a critical factor. It was later demonstrated that increased knowledge about diagnosis and treatment of depression among the general practitioners at Gotland resulted in decreased morbidity and decreased mortality [51–55].

Possible mechanisms for increased long-term morbidity and mortality

It has been known for a long time [56] that depressed patients tend to have increased cortisol in serum and reduced diurnal variation in the cortisol secretion. Perceived stress, resulting in a uncontrollable defeat reaction, has been shown to be followed by specific endocrine abnormalities. In a series of studies, longstanding uncontrollable perceived stress has been linked to a specific endocrine syndrome known as the metabolic syndrome, characterised by a specific type of obesity with increased Waist to Hip Ratio (WHR), increased cortisol secretion and decreased secretion of sex hormones, growth hormones and increased mortality in diabetes, and cardiovascular diseases [57] (Fig. 6). There seem to be many similarities between longstanding depressive disorders with long standing hypercortisolemia and the metabolic syndrome.

In depressed patients non-suppression of cortisol secretion by dexamethasone has been demonstrated to be a common finding [58], although the test is not as specific for melancholia as first believed [59]. In the metabolic syndrome, a sensitisation of the hypothalamo-pituitary-adrenal (HPA) axis [57] has been demonstrated. The activity of the HPA axis is regulated by central glucocorticoid receptors whose activity can be tested by the administration of exogenous glucocorticoids, which normally inhibit cortisol secretion. The sensitivity of the system has been tested by means of a dose response study of the inhibition of cortisol secretion by dexamethasone [60]. The inhibition of the cortisol secretion at 0.5 mg dexamethasone

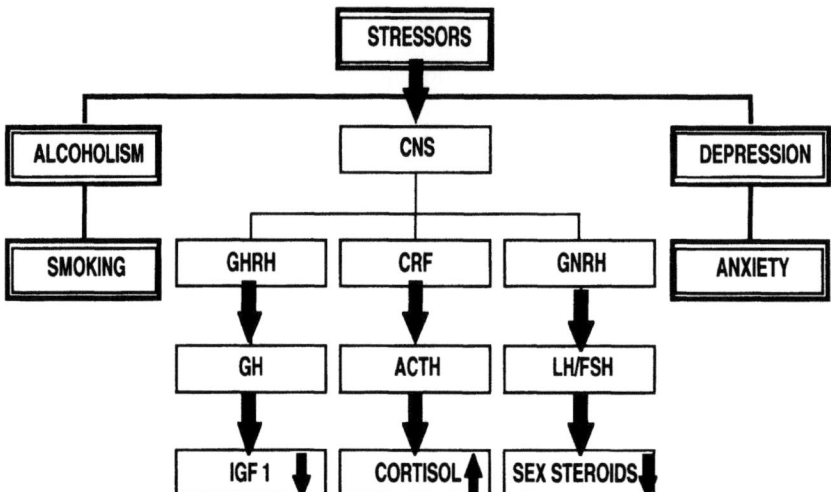

Figure 6. Possible mechanisms for increased long-term morbidity and mortality of the insulin-resistance syndrome [64] or the metabolic syndrome [57].

was negatively correlated with the WHR and was significantly lower in men with a WHR of > 1.0 than in men with a WHR < 1.0. Thus men with an elevated WHR experienced a decrease in the inhibition of cortisol secretion by dexamethasone [60]. It has also been shown that women with visceral fat accumulation and an increased WHR have elevated cortisol secretion and an increased sensitivity along the HPA axis [61].

Stress related cortisol secretion has also been associated with concentrations of insulin-like growth factor 1 (IGF-1), testosterone, total and low density lipoprotein (LDL) cholesterol, diastolic fasting insulin, glucose, triglycerides and diastolic blood pressure [62]. Furthermore, the stress-related cortisol secretion is strongly and consistently related to obesity factors (Body Mass Index, BMI, and WHR), metabolic variables (insulin, glucose, triglycerides, total and LDL cholesterol) and hemodynamic variables (systolic and diastolic blood pressure and heart rate) [62]. An increased WHR, indicating intraabdominal fat masses, increased cortisol secretion and insulin resistance are regarded as powerful risk factors for cardiovascular disease, diabetes and stroke [57].

In middle-aged men, the WHR has been significantly related to the use of anxiolytics, hypnotics, antidepressive drugs, degree of melancholy, life satisfaction, difficulties in sleeping and dyspepsia [63]. Patients with abdominal obesity and high WHR were found to have low high-density lipoprotein cholesterol (HDLC), high triglycerides, high insulin and C-peptide levels and high ACTH response to glucose [64]. Thus, it seems as if abdominal obesity is associated with a subtle central adrenal insufficiency which also affects insulin and lipoprotein metabolism, known as the insulin resistance syndrome [64].

In conclusion, it has been suggested that depressive disorders, known to be long-lasting and recurrent, affecting approximately 20% of the lifetime, may act as a long-lasting negative stressor of the defeat type. The long-lasting depressive episodes are often associated with psychosocial and socioeconomic handicaps, alcohol consumption and smoking. During such episodes there is high cortisol secretion, decreased sensitivity to dexamethasone, accumulation of intraabdominal visceral fat masses, insulin resistance and other risk factors dependent on the hyperinsulinemia following insulin resistance. The low sex steroid and growth hormone secretions might be secondary to the hypersensitive HPA axis [57]. In the end there is increased risk of cardiovascular disease, diabetes and stroke, resulting in increased mortality as earlier described in depressed patients.

References

1 Hylan T (1997) Comparisons of costs in antidepressant therapy. Eli Lilly Company, Arlanda, Sweden
2 Bingefors K, Isacson D, von Knorring L, Smedby B (1995) Prescription drug and health care use among Swedish patients treated with antidepressants. *Annals Pharmacotherapy* 29: 566–572
3 Ekselius L (1994) Personality disorders in the DSM-III-R. Uppsala University, Uppsala
4 Perris C (1966) A study of bipolar (manic-depressive) and unipolar recurrent depressive psychoses X. Mortality, suicide and life-cycles. *Acta Psychiatrica Scandinavica* 42: 172–183
5 Bingefors K, Isacson D, von Knorring L, Smedby B, Wicknertz K (1996) Antidepressant-treated patients in ambulatory care. Mortality during a nine-year period after first treatment. *Brit J Psychiatry* 169: 647–654
6 Canals J, Domenech E, Carbajo G, Blade J (1997) Prevalence of DSM-III-R and ICD-10 psychiatric disorders in a Spanish population of 18-year-olds *Acta Psychiatrica Scandinavica* 96: 287–294
7 Association AP (1987) *Diagnostic and statistical manual of mental disorders*, 3rd edition, revised. APA, Washington, DC
8 World Health Organization (1993) The ICD-10 classification of mental and behavioural disorders: diagnostic criteria for research. WHO, Geneva
9 Gorman J (1996/1997) Comorbid depression and anxiety spectrum disorders. *Depression Anxiety* 4: 160–168
10 Thapar A, McGuffin P (1997) Anxiety and depressive symptoms in childhood – A genetic study of comorbidity. *J Child Psych Psychiatry Allied Discip* 38: 651–656
11 Rodney J, Prior N, Cooper B, Theodoros M, Browning J, Steinberg B, Evans L (1997) The comorbidity of anxiety and depression. *Australian New Zealand J Psych* 31: 700–703
12 Sherbourne C, Wells K (1997) Course of depression in patients with comorbid anxiety disorders. *J Affective Disorders* 43: 245–250
13 Angst J (1997) Depression and anxiety: Implications for nosology, course, and treatment. *J Clin Psych* 58: 3–5
14 Kendler K, Davis C, Kessler R (1997) The familial aggregation of common psychiatric and substance use disorders in the National Comorbidity survey: A family history study. *Brit J Psychiatry* 170: 541–548
15 Schrader G (1997) Subjective and objective assessments of medical comorbidity in chronic depression. *Psychother Psychosom* 66: 258–260
16 American Psychiatric Association (1994) *Diagnostic and statistical manual of mental disorders*. American Psychiatric Association, Washington, DC
17 Corruble E, Ginestet D, Guelfi J (1996) Comorbidity of personality disorders and unipolar major depression: A review. *J Affective Disorders* 37: 157–170

18 Ekselius L, von Knorring L , Eberhard G (1997) A double-blind multicenter trial comparing sertraline and citalopram in patients with major depression treated in general practice. *Internat Clin Psychopharma* 12: 323–331
19 Widiger T (1989) The categorical distinction between personality and affective disorders. *J Personality Dis* 3: 77–91
20 Farmer R, Nelson-Grey R (1990) Personality disorders and depression: Hypothetical relations, empirical findings, and methodological considerations. *Clin Psych Rev* 10: 453–476
21 Kiloh L, Ball J, Garside R (1962) Prognostic factors in the treatment of depressive states with imipramine. *Brit Med J* 1: 1225–1227
22 Pfohl B, Stangl D, Zimmerman M (1984) The implications of DSM-III disorders for patients with major depression. *J Affective Disorders* 7: 309–318
23 Black D, Bell S, Hulbert J, Nasrallah A (1988) The importance of Axis II in patients with major depression. *J Affective Disorders* 14: 115–122
24 Reich J (1990) Effect of DSM-III personality disorders on outcome of tricyclic antidepressant treated nonpsychotic outpatients with major or minor depressive disorders. *Psychiatry Res* 32: 175–181
25 Akiskal H, Bitar A, Puzantian V, Rosenthal T, Walker P (1978) The nosological status of neurotic depression. *Arch General Psych* 35: 756–766
26 Thompson L, Gallagher D, Czirr R (1988) Personality disorder and outcome in the treatment of late life depression. *J Geriatric Psych* 21: 133–150
27 Ansseau M, Troisfonatines B, Papart P, von Frenckel R (1991) Compulsive personality as predictor of response to sertonergic antidepressants. *Brit Med J* 303: 760–761
28 Fava M, Bouffides E, Pava J, McCarthy M, Steingard R, Rosenbaum J (1994) Personality disorder comorbidity with major depression and response to fluoxetine treatment. *Psychotherapy Psychosom* 62: 160–167
29 Fava M, Uebelacker L, Alpert J, Nirenberg A, Pava J, Rosenbaum J (1997) Major depressive subtypes and treatment response. *Biological Psych* 42: 568–576
30 Patience D, McGuire R, Scott A, Freeman C (1995) The Edinburgh primary care depression study: Personality disorder and outcome. *Brit J Psych* 167: 324–330
31 Chynoweth R, Tonge J, Armstrong J (1980) Suicide in Brisbane – a retrospective psychosocial study. *Australian New Zealand J Psych* 14: 37–45
32 Lesage A, Boyer R, Gruenberg F, Vanier C, Morissette R, Menard-Buteau C, Loyer M (1994) Suicide and mental disorders: a case control study of young men. *Am J Psych* 151: 1063–1068
33 Cheng A, Mann A, Chan-KA (1997) Personality disorder and suicide: A case-control study. *Brit J Psych* 170: 441–446
34 Silverstone P (1996) Prevalence of psychiatric disorders in psychiatric inpatients. *J Nerv Ment Dis* 1996: 43–51
35 Endicott J (1984) Measurement of depression in patients with cancer. *Cancer* 53: 2243–2248
36 Depression Guideline Panel (1993) *Depression in primary care. Volume 1. Detection and diagnosis. Clinical practical guideline.* No. 1. U.S. Department of Health and Human Services, Public Health Services, Agency for Health Policy and Research, Rockville, MD
37 Perris C (1966) A study of bipolar (manic-depressive) and recurrent depressive psychoses. *Acta Psychiatrica Scandinavica Suppl.* 194
38 Keitner G, Ryan C, Miller I, Kohn R, Epstein N (1991) 12-month outcome of patients with major depression and comorbid psychiatric or medical illness (compund depression). *Am J Psychiatry* 148: 45–50
39 Gregory R, Jimerson D, Walton B, Daley J, Paulsen R (1992) Pharmacotherapy of depression in the medically ill: Directions for future research. *Gen Hosp Psychiatry* 14: 36–42
40 Widmer R, Cadoret R (1983) Depression: The great imitator in family practice. *J Fam Pract* 17: 485–505
41 Wilkinson G, Smeeton N, Skuse D, Fry J (1988) Conultation for physical illnesses by patients diagnosed and treated for psychiatric disorders by general practitioner: 20 year follow up study. *Br Med J* 297: 776–778
42 Bingefors K, Isacson D, von Knorring L, Smedby B, Ekselius L, Kupper L (1996) Antidepressant-treated patients in ambulatory care. Long-term use of non-psychotropic and psychotropic drugs. *Brit J Psychiatry* 168: 292–298
43 Sims A (1988) The mortality associated with depression. *Int Clin Psychopharm* 3: 1–13

44 Carney R, Rich M, Freedland K, Saini J, deVelde A, Simeone C (1988) Major depressive disorder predicts cardiac events in patients with coronary artery disease. *Psychosom Med* 50: 627–633
45 Frasure-Smith N, Lesperance F, Talajic M (1993) Depression following myocardial infarction. Impact on 6-month survival. *JAMA* 270: 1819–1825
46 Bruce M, Leaf P, Rozal G, Florio L, Hoff R (1994) Psychiatric status and 9-year mortality data in the New Haven Epidemiological Catchment Area Study. *Am J Psychiatry* 151: 716–721
47 Harris E, Barraclough B (1998) Excess mortality of mental disorder. *Br J Psychiatry* 173: 11–53
48 Editorial (1992) Depression and suicide: are they preventable? *Lancet* 340: 700–701
49 Coppen A, Standish-Berry H, Bailey J, Houston G, Silcocks P, Hermon C (1991) Does lithium reduce the mortality of recurrent mood disorders? *J Affective Dis* 23: 1–7
50 Muller-Oerlinghausen B, Ahrens B, Volk J et al (1991) Reduced mortality of manic-depressive patients in long-term lithium treatment: an international collaborative study by IGSLI. *Psychiatry Res* 36: 329–331
51 Rutz W, Wålinder J, Eberhard G, Holmberg G, von Knorring A-L, von Knorring L, Wistedt B, Åberg-Wistedt A (1989) An educational program on depressive disorders for general practitioners on Gotland: Background and evaluation. *Acta Psychiatrica Scandinavica* 79: 19–26
52 Rutz W, von Knorring L, Wålinder J, Wistedt B (1990) Effect of an educational program for general practitioners on Gotland on the pattern of prescription of psychotropic drugs. *Acta Psychiatrica Scandinavica* 82: 399–403
53 Rutz W, von Knorring L, Wålinder J (1992) Long-term effects of an educational program for general practitioners given by the Swedish Committee for Prevention and Treatment of Depression. *Acta Psychiatrica Scandinavica* 85: 83–88
54 Rutz W, Carlsson P, von Knorring L, Wålinder J (1992) Cost-benefit analysis of an educational program for general practitioners by the Swedish Committee for the Prevention and Treatment of Depression. *Acta Psychiatrica Scandinavica* 85: 457–464
55 Rutz W, von Knorring L, Pihlgren H, Rihmer Z, Wålinder J (1995) An educational project on depression and its consequences: is frequency of major depression among Swedish men underrated, resulting in high suicidality? *Primary Care Psychiatry* 1: 59–63
56 Sacher E, Roffwarg H, Gruen P (1976) Neuroendocrine studies of depressive illness. *Pharmacopsychiatry* 9: 11–17
57 Björntorp P (1997) Hormonal control of regional fat distribution. Hum Reprod 12: 21–25
58 Carroll B (1982) The dexamethasone suppression test for melancholia. *Br J Psychiatry* 140: 292–304
59 Hällström T, Samuelsson S, Balldin J (1983) Abnormal dexamethasone suppression test in normal females. *Br J Psychiatry* 142: 489–497
60 Ljung T, Andersson B, Bengtsson B, Björntorp P, Mårin P (1996) Inhibition of cortisol secretion by dexamethasone in relation to body fat distribution: a dose-response study. *Obes Res* 4: 277–282
61 Mårin P, Darin N, Amemiya T, Andersson B, Jern S, Björntorp P (1992) Cortisol secretion in relation to body fat distribution in obese premenopausal women. *Metabolism* 41: 882–886
62 Rosmond R, Dallman M, Björntorp P (1998) Stress-related cortisol secretion in men: relationships with abdominal obesity and endocrine, metabolic and hemodynamic abnormalities. *J Clin Endocrin Metab* 83: 1853–1859
63 Rosmond R, Lapidus L, Mårin P, Björntorp P (1996) Mental distress, obesity and body fat distribution in middle-aged men. *Obes Res* 4: 245–252
64 Hautanen A, Adlercreutz H (1993) Altered adrenocorticotropin and cortisol secretion in abdominal obesity: implications for the insulin resistence syndrome. *J Intern Med* 234: 461–469
65 Bingefors K (1996) Antidepressant-treated patients. Population-based longitudinal studies. Uppsala University, Uppsala

Do antidepressants really take several weeks to show effect?

Jules Angst and Hans H. Stassen

Psychiatric University Hospital Zurich, Research Department, P.O. Box 68, CH-8029 Zurich, Switzerland

Introduction

Conventional repeated measurements analysis of variance suggests that response to antidepressants is slow, and that differences between antidepressants and placebo do not normally reach statistical significance until the 3rd or 4th week of treatment. This statistical "response lag" has finally led to the "delayed onset of action" hypothesis, although response-to-treatment and onset-of-action represent two principally different concepts that relate to different aspects of efficacy of antidepressant drug therapy. While response-to-treatment has its main focus on the proportion of patients in whom antidepressants induce a clinically relevant change, onset-of-action refers to the speed at which symptoms reduce under antidepressants in comparison to placebo. Standard drug trial procedures tend to obscure differences in onset of action across drugs, as well as across patients, because such procedures rely upon mean depression scores derived from studies which amalgamate separate depressive subgroups, rapid and slow remitters, as well as partial remitters, non-remitters and premature withdrawals (Parker, 1996a; Stassen et al., 1998).

Clinical reports that deal with the time-course of recovery from depression under antidepressants and placebo suggest that there is no wide-spread agreement about the delayed onset of the therapeutic effect of antidepressants. And, most surprisingly, patients demonstrating clinical improvement within the first 2 weeks of therapy have surfaced in almost every antidepressant drug trial. Katz et al. (1987), for example, reported about their drug trial comparing amitriptyline with imipramine: "It is clear from this study that when drug treatment is going to be effective, significant positive changes in behavior occur within the first week of treatment. The earliest changes associated with recovery are in the levels of disturbed affect and cognitive functioning". Similarly, Khan et al. (1989) found in their investigation of onset of response in relation to outcome under imipramine and placebo: "The early onset response was seen in a subgroup of patients receiving either imipramine or placebo and appeared to be independent of treatment assignment. Furthermore, the early onset response predicted outcome for the duration of the trial and was not selective as defined by specific changes in subscales measuring insomnia, anxiety or endogenous features". These findings appear to be consistent with clinical practice, as described, for example, by Grey (1993) in his observation of the therapeutic effects of fluoxetine: "The onset of response in my patients has most frequently been within 12–36 hours and quite dramatic. Of the last 18 patients for whom I have prescribed fluoxetine, 11 had an immediate effect".

However, Quitkin et al. (1987) pointed out "that much of the improvement observed shortly after initiating treatment may be a placebo effect or attributed to improved sleep", and "that abrupt, nonpersistent improvement may be an indicator of placebo response". In consequence, the authors concluded in their study comparing phenelzine and imipramine with placebo: "There was little evidence for a true drug effect within the first two weeks of treatment", thus raising the question of what makes true drug responders distinguishable from placebo responders, and how effective antidepressants really are. Since answers to these issues would be of major theoretical and practical relevance, an increasing number of investigations into the onset and time-course of improvement under various kinds of antidepressant treatment, mostly in comparison with placebo, have been published during recent years (Angst et al., 1993; Quitkin et al., 1993; Stassen et al., 1993; Tollefson and Holman, 1994; Montgomery, 1995; Möller et al., 1996; Parker 1996a, b; Quitkin et al., 1996; Katz et al., 1997a, b; Stassen et al., 1997, 1998; Müller and Möller, 1998).

The current data available on the effects of antidepressants remain nonetheless puzzling, despite the many attempts to detail the trajectory of recovery from depression (Parker, 1996b): (1) antidepressants that differ greatly in their biochemical and pharmacological actions have virtually the same efficacy in terms of the proportion of patients in whom they induce a therapeutic response; (2) a large proportion of patients (35 to 45%) have a refractory depression, which is resistant to all drug treatments; (3) a large proportion of patients (typically 30 to 40% among moderately to severely depressed patients) are placebo responders, and (4) under the assumption that placebo response rates are equal for all treatment modalities, only 20 to 25% of patients are, on average, "true" drug responders.

Undoubtedly, the reliable identification of early-, late-, partial- and non-remitters would contribute to a better understanding of these issues. Evidence from our own studies, now replicated across several different antidepressant drug classes and placebo, suggests that the therapeutic qualities of antidepressants may not lie in their suppression of symptoms, but rather in their ability to change a complex regulatory system at one or the other point and to convert a percentage of nonresponders to responders, triggering and maintaining the conditions necessary for improvement.

Onset of action as assessed by refined methods

In recent years survival-analytical methods have become a valuable extension of standard drug trial procedures with respect to the timing issues of antidepressant drug response (Greenhouse et al., 1989; Stassen et al., 1993; Lavori et al., 1994, 1996). Unlike the traditionally accepted statistical approach to evaluating treatment-induced change, where psychopathology scores (or changes thereof) are analysed as a function of time, survival-analytical methods treat time as a function of symptom reduction. This has the advantage that variations around the pre-specified, cross-sectional observation days (typically ± 2 days) can be incorporated into the model, thus providing (1) a better temporal resolution with respect to the individual course of recovery from depression, (2) a way to distinguish between early and late responders, and (3) the possibility to assess the predictive value of early improvement with respect to later outcome. Additionally, these methods allow one to take into account that treatment response, or nonresponse, has not been observed in a subgroup of patients due to premature withdrawal or the limited observation time.

Because of the lack of knowledge about the specific effects of antidepressants, the onset of action of these drugs can only be assessed indirectly by an analysis of the time-course of improvement. To this end, the statistical distribution of the time spans to onset of improvement yields important insights into the nature of drug effects. If a distinct mode of action of antidepressants exists, which, according to the hypothesis of delayed onset of action, is expected to emerge after 2 to 3 weeks of treatment, a clearly visible peak in the distribution of time spans to onset of improvement should be present somewhere after day 14. And, if one drug had a faster onset of action than another, this should be detectable through differences in the respective distribution curves.

In order to detail the time-course of improvement for different classes of antidepressants, we carried out a series of meta-analyses using the above method of approach: a meta-analysis of a double-blind drug trial comparing the tricyclic (TCA) amitriptyline (n = 120) and the highly selective noradrenaline reuptake inhibitor oxaprotiline (n = 120) with placebo (n = 189), a meta-analysis of a double-blind drug trial comparing the tricyclic imipramine (n = 506) and the reversible and selective MAO-A inhibitor moclobemide (n = 580) with placebo (n = 191), and a meta-analysis of a double-blind drug trial comparing the selective serotonin reuptake inhibitor (SSRI) fluoxetine (n = 440) with the reversible and selective MAO-A inhibitor moclobemide (n = 437). The meta-analysis approach was chosen because the results from studies that compare efficacy and time characteristics of antidepressants with different mechanisms of action are difficult to interpret unless very large numbers of patients are included.

The patient samples consisted of moderately to severely depressed patients, diagnosed according to DSM-III criteria as suffering from major depressive disorders. To be included, patients were required to demonstrate a baseline score (end of wash-out period) of at least 15 on the 17-item Hamilton Depression Scale. Efficacy and safety assessments were carried out at entry into trial, at day-3 and day-7 of the trial and, subsequently, at weekly intervals (typically 5 weeks). Measures of efficacy were provided by the Hamilton Depression (HAMD) and the Clinical Global Improvement (CGI) rating scales. In these drug trials, cross-sectional analyses generally demonstrated a significant advantage after week three in those patients receiving active drug. Our interest in these meta-analyses was focused on (1) the distribution of time spans to onset of improvement, (2) the proportion of patients who showed improvement within the first 2 weeks of treatment, (3) the time-course of improvement for each individual patient throughout the observation period, and (4) the predictive value of early improvement with respect to later outcome. To this end, we distinguished between "improvement" (20–40% HAMD baseline score reduction without subsequent deterioration beyond 15% of achieved HAMD score) and "response to treatment" (50–60% HAMD baseline score reduction without subsequent deterioration beyond 15% of achieved HAMD score).

As expected, the differences in the total number of improvers/responders were highly significant when comparing active drugs and placebo, whereas the differences within active drugs did not always reach significance. Yet unexpectedly, among improvers we found the distributions of time spans to onset of improvement to be virtually independent of treatment modality, as reflected, for example, by the cumulative percentage rates of improvers that turned out to be virtually the same across the treatment groups when regarded as a function of time (Tabs. 1, 2, 3). The obvious differences between the three tables with respect to the "null point" of the time scale are

Table 1. Cumulative rates of improvers in patients receiving imipramine, moclobemide and placebo, as derived by survival-analytical methods. The onset of improvement was defined by a 20% reduction of HAMD baseline score without subsequent deterioration. Percentages for each time-point are calculated for each treatment group with respect to the effective total number of improvers, excluding nonimprovers (Stassen et al., 1997)

Day	Cumulative percentage rates of improvers (%)		
	Imipramine $n = 419$	Moclobemide $n = 453$	Placebo $n = 111$
3	15.4	17.8	23.3
7	55.1	53.7	54.3
10	61.9	61.2	60.3
14	81.3	80.1	77.6
21	93.5	92.1	89.7
28	97.9	97.0	95.7
Percentage of total sample	83%	78%	58%

Table 2. Cumulative rates of improvers in patients receiving oxaprotiline, amitriptyline and placebo, as derived by survival-analytical methods. The onset of improvement was defined by a 20% reduction of HAMD baseline score without subsequent deterioration. Percentages for each time-point are calculated for each treatment group with respect to the effective total number of improvers, excluding nonimprovers (Stassen et al., 1997)

Day	Cumulative percentage rates of improvers (%)		
	Oxaprotiline n = 90	Amitriptyline n = 102	Placebo n = 120
3	0.0	0.0	0.0
7	3.3	1.9	1.1
10	22.0	24.3	20.4
14	48.4	55.1	53.8
21	73.6	78.5	82.8
28	95.6	99.1	92.5
Percentage of total sample	75%	85%	64%

artifactual, due to differences in the availability of data on the wash-out period which were included in the analysis where possible in order to study the time course of improvement at the earliest possible stage.

The cumulative rates remained the same across the treatment groups, even when the improvement criterion was successively tightened up from 20 to 60%, thus suggesting that the time-course of improvement among responders was independent of the treatment modality. In particular, no treatment-dependent time lag showed up in the onset of improvement.

In other words, our analyses of various classes of antidepressants and placebo revealed a continuous distribution of time spans to onset of improvement with a maximum between day-10 and day-14, but no indication

Table 3. Cumulative rates of improvers in patients receiving fluoxetine and moclobemide, as derived by survival-analytical methods. The onset of improvement was defined by a 20% reduction of HAMD baseline score without subsequent deterioration. Percentages for each time-point are calculated for each treatment group with respect to the effective total number of improvers, excluding nonimprovers (Stassen et al., 1998)

Day	Cumulative percentage rates of improvers (%)	
	Fluoxetine n = 345	Moclobemide n = 348
3	3.6	3.0
7	33.7	33.8
10	43.5	43.7
14	67.5	66.8
21	79.0	81.9
28	94.7	94.8
Percentage of total sample	78%	80%

of a distinct drug effect with respect to the time characteristics of recovery from depression. Since all HAMD items appeared to contribute to score reduction at the early stages of therapy, the early onset of action of antidepressants cannot be explained as an artifact of rating instruments. Rather, with respect to drug- or placebo-induced change, our analyses demonstrated that, as a rule, onset of improvement occurred in more than 50% of cases within the first 2 weeks of treatment, and not more than 5–10% of cases showed improvement after the 4th week. Response to treatment was observed in 25% of cases within the first 2 weeks of treatment. These figures, however, apply only to a sufficiently representative population of depressive patients, since response to treatment is influenced by many other factors. For example, we found response rates in patients pretreated with TCAs to be typically 10% lower, and patients whose previous course of illness was characterized by short episodes exhibited higher response rates than patients with longer previous episodes. Moreover, placebo response rates were typically 35 to 40% for mild to moderate depression (HAMD-17 baseline score ≤ 27), whereas in severe cases (HAMD-17 baseline score > 27), the placebo rate fell to 20%.

The early onset of improvement was highly predictive of later outcome. For example, 79.6% of imipramine-treated patients who showed improvement within the first 2 weeks of treatment became treatment responders, i.e. exhibited a 50% HAMD baseline score reduction without subsequent deterioration. The respective rates were $> 75\%$ for all other treatment modalities, except for fluoxetine (70%) and placebo (61–65%). Conversely, $\geq 75\%$ of all treatment and placebo responders exhibited an onset of improvement within the first 2 weeks of treatment.

In summary, differences in efficacy between active drugs and placebo were reflected only by the total number of improvers and responders, whereas the time-course of improvement turned out to be independent of treatment modality. Effective antidepressants appeared to merely trigger and maintain the conditions necessary for improvement. Once triggered, the recovery followed its natural course identical to that of spontaneous remission, as observed under placebo treatment. Consequently, depressed patients may have a biological predisposition to be a responder or nonresponder. This viewpoint has gained strong support by recent pharmacogenetic findings which indicate that there exist not only ethnic differences but also marked inter-individual variation within ethnic groups with respect to basic properties of the monoaminergic systems and drug metabolism.

Biological predisposition to drug response

Over the past decades, the therapeutic effect of various types of antidepressant drugs has been successfully conceptualized on the basis of monoaminergic system modifications. Although achieved via different mechanisms, common to these modifications is the enhancement in neurotransmitter signal transfer, which suggests that an inadequate concentration of biogenic amines at synapses and in neurons, for whatever reason, is involved in the pathogenesis of depression. However, empirical data from numerous efficacy studies indicate that the effects of antidepressants are unspecific and virtually independent of the biochemical mode, as well as of the primary site of action within the monoaminergic systems. The apparent nonspecificity of antidepressants with respect to inducing changes of depressive symptomatology is further supported by pharmacogenetic studies that revealed ethnic variation in dose, blood level, and side effects for all psychotropic drugs, although response rates do not show ethnic differences (Ruiz et al., 1996; Bertilsson et al., 1997; Masimirembwa and Hasler, 1997; Meyer and Zanger, 1997; Sachse et al., 1997). Moreover, recent molecular-genetic studies of the cytochrome P450 enzyme system, that is involved in the metabolism of endogenous substances, drugs and xenobiotics, demonstrated that there exists considerable genotypic interindividual variation within populations (Coutts, 1994; DeVane, 1994; Daly, 1995; Eichelbaum and Evert, 1996). However, correlations between plasma levels and efficacy are generally poor as demonstrated, for example, by the fact that the steady-state plasma concentrations of fluoxetine or norfluoxetine do not appear to be related to relapse and long-term therapeutic outcome (Brunswick et al., 1998). Accordingly, the benefit of genotyping patients prior to drug therapy (in order to select an adequate drug and its dosis regimen on a biological basis) remains unclear.

On the other hand, genotypic variation explains less than 50% of observed phenotypic variation, as suggested by twin or adoption studies (Wender et al., 1986; Kendler et al., 1993). Known environmental factors that influence response to antidepressants encompass diet, coffee, tobacco, pollution, and co-medication, among others. With respect to co-medication, the pretreatment with herbal preparations has gained increasing attention during recent years, since a steadily increasing proportion of the general population sees an alternative health care provider for alternative medical treatments, and the most common reason given for herb ingestion (such as St. John's Wort) is to treat symptoms of depression, to reduce anxiety or to improve sleep (Ware et al., 1998).

Another difficulty in dealing with the time characteristics of antidepressant drug response originates from diagnosis. Any patient sample randomly compiled under the "major depression" criterion is likely to include the full spectrum of individual response characteristics, rapid and slow remitters, as well as partial remitters and nonremitters. In other words, current diagnostic entities do not appear to be useful for research purposes and "may even border on insignificance as far as clinical practice is concerned" (van Praag, 1998). In fact, little is known about the aetiology of depression, and diagnostic differentiation based on biological markers and objective laboratory methods remains to be established.

Consequences

Analyses based on improved methods revealed no evidence for the delayed onset of action of various antidepressants with large biochemical and pharmacological differences when compared to placebo, thus underlining the viewpoint put forward by Angst (1978) and Willner (1989), who described delayed onset of action under antidepressants as *"something of a myth"* due to drawbacks of research methodology. If effective antidepressants do not have a delayed onset of action, what clinical implications does this have? Since the typical response of patients treated with antidepressants is immediate and early improvement is predictive of later outcome, probability-based guidelines can be developed for clinicians that include estimates of false-positives[1], as well as of false-negatives[1], at each time of treatment. Based on current data, we would like to make the following recommendations: (1) More than 70% of patients showing at least slight improvement during the first 14 days of antidepressant therapy will become treatment responders. In this regard, "slight" means a 20% decrease on a depression rating scale, even if the patient does not "feel" better. If this kind of decrease is observed, antidepressant therapy should be continued. (2) On average, no more than 10% of patients who do not show any improvement within the first 3 weeks of treatment will become responders. In such cases, a change to an antidepressant of a different class should be considered by the end of week 3. However, any change in medication must be decided upon individually, since this decision depends on many other factors, such as the severity of the patient's symptoms, number and duration of previous episodes, or previous treatment with tricyclics. (3) The major differences between antidepressants relate to their side-effect profiles and general toxicity, but much less to their overall efficacy or their onset of action. It is important to note, however, that our analyses have dealt entirely with acute depression and therefore, results do not apply to chronic cases. Moreover, differences between active compounds and placebo become marginal in mild depression (Angst et al. 1993).

Given current response rates of antidepressants together with the distinct heterogeneity of individual time-courses of recovery under antidepressant treatment, future research must acknowledge that mechanisms different from those related to the monoaminergic systems are likely to be involved in the pathogenesis of depression. Indeed, hormonal alterations in depressed patients have long been evident. Among these alterations the dysregulation of the hypothalamic-pituitary-adrenal (HPA) axis is most intriguing, since chronic activation of this axis has been hypothesized to produce depressive symptoms. There may be other such systems (Kramer

---

[1] The term "false-positives" refers to patients falsely treated with antidepressants (no benefit), while that of "false-negatives" to patients falsely not treated with antidepressants (would benefit from treatment).

et al. 1998; Plotsky et al. 1998; Holsboer 1999). More work on the neurobiology of depression is clearly indicated to make treatment more specific, and thus more effective with respect to the proportion of patients in whom a therapeutic response is induced, as well as to the time characteristics of response.

Acknowledgements
This article summarizes findings previously published in scientific journals.

References

Angst J (1978) Time-lag of antidepressant effect. In: *Symposia Medica Hoechst 13*, Symposium Rome 1977: Depressive Disorders. Stuttgart, Schattauer, 468

Angst J, Delini-Stula A, Stassen HH (1993) Is a cutoff score a suitable measure for treament outcome? A methodological meta-analysis. *Human Psychopharmacology* 8: 311–317

Bertilsson L, Dahl ML, Tybring G (1997) Pharmacogenetics of antidepressants: clinical aspects. *Acta Psychiatr Scand* 391: 14–21

Brunswick DJ, Amsterdam JD, Fawcett J, Quitkin F, Reimherr F, Rosenbaum J, Beasley C (1998) Relation of steady-state fluoxetine plasma levels and relapse-prevention outcome. NCDEU 38th Annual Meeting, Abstracts 180

Coutts RT (1994) Polymorphism in the metabolism of drugs, including antidepressant drugs: comments on phenotyping. *J Psychiatry Neurosci* 19 (1): 30–44

DeVane CL (1994) Pharmacogenetics and drug metabolism of newer antidepressant agents. *J Clin Psychiatry* 55: 38–45

Daly AK (1995) Molecular basis of polymorphic drug metabolism. *J Mol Med* 73 (11): 539–553

Eichelbaum M, Evert B (1996) Influence of pharmacogenetics on drug disposition and response. *Clin Exp Pharmacol Physiol* 23 (10–11): 983–985

Greenhouse JB, Stangl D, Bromberg J (1989) An introduction to survival analysis: statistical methods for analysis of clinical trial data. *J Cons Clin Psychology* 57: 536–544

Grey P (1993) Fluoxetine and onset of its therapeutic effect. *Am J Psychiatry* 150: 984

Holsboer F (1999) The rationale for corticotropin-releasing hormone receptor (CRH-R) antagonists to treat depression and anxiety. *J Psychiatric Res* 33(3): 181–214

Katz MM, Koslow SH, Maas JW, Frazer A, Bowden CL, Casper R, Croughan J, Kocsis J, Redmond E jr (1987) The timing, specificity and clinical prediction of tricycling drug effects in depression. *Psychol Med* 17: 297–309

Katz MM, Koslow SH, Frazer A (1997a) Onset of antidepressant activity: reexamining the structure of depression and multiple actions of drugs. *Depression and Anxiety* 4: 257–267

Katz MM, Bowden C, Stokes P, Casper R, Frazer A, Koslow SH, Kocsis J, Secunda S, Swann A, Berman N (1997b) Can the effects of antidepressants be observed in the first two weeks of treatment? *Neuropsychopharmacology* 17: 110–111

Kendler KS, Pedersen N, Johnson L, Neale MC, Mathé AA (1993) A pilot Swedish twin study of affective illness, including hospital- and population-ascertained subsamples. *Arch Gen Psychiatry* 50: 699–706

Khan A, Cohen S, Dager S, Avery DH, Dunner DL (1989) Onset of response in relation to outcome in depressed outpatients with placebo and imipramine. *J Affect Disord* 17: 33–38

Kramer MS, Cutler N, Feighner J, Shrivastava R, Carman J, Sramek JJ, Reines SC, Liu G, Snavely D, Wyatt-Knowles E et al. (1998) Distinct mechanism for antidepressant activity by blockade of central substance P receptors. *Science* 281: 1640–1645

Lavori PW, Dawson R, Mueller TI (1994) Causal estimation of time-varying treatment effects in observational studies: application to depressive disorder. *Statistics in Medicine* 13: 1089–1100

Lavori PW, Dawson R, Mueller TI, Warshaw M, Swartz A, Leon A (1996) Analysis of course of psychopathology: transitions among states of health and illness. *Int J Meth Psychiat Res* 6: 321–334

Masimirembwa CM, Hasler JA (1997) Genetic polymorphism of drug metabolising enzymes in African populations: implications for the use of neuroleptics and antidepressants. *Brain Res Bull* 44 (5): 561–571

Meyer UA, Zanger UM (1997) Molecular mechanisms of genetic polymorphisms of drug metabolism. *Annu Rev Pharmacol Toxicol* 37: 269–296

Möller H-J, Müller H, Volz H-P (1996) How to assess the onset of antidepressant effect: comparison of global ratings and findings based on depression scales. *Pharmacopsychiatry* 29: 57–62

Montgomery SA (1995) Rapid onset of action of venlafaxine. *Int Clin Psychopharmacol* 10, Suppl. 2: 21–27

Müller H, Möller H-J (1998) Methodological problems in the estimation of the onset of the antidepressant effect. *J Aff Disorder* 48: 15–23

Parker G (1996a) On brightening up: triggers and trajectories to recovery from depression. *Br J Psychiatry* 168: 263–264

Parker G (1996b) Recovery from depression: triggers and time patterns. *Australian and New Zealand J Psychiatry* 30: 442–444

Plotsky PM, Owens MJ, Nemeroff CB (1998) Psychoendocrinology of depression: hypothalamic-pituitary-adrenal axis. *Psychatr Clin North Am* 21: 293–307

Quitkin FM, Rabkin JD, Markowitz JM, Stewart JW, McGrath PJ, Harrison W (1987) Use of pattern analysis to identify true drug response. *Arch Gen Psychiatry* 44: 259–264

Quitkin FM, Stewart JW, McGrath PJ, Nunes E, Ocepek-Welikson K, Tricamo E, Rabkin JG, Klein DF (1993) Further evidence that a placebo response to antidepressants can be identified. *Am J Psychiatry* 150: 566–570

Quitkin FM, McGrath PJ, Stewart JW, Taylor BP, Klein DF (1996) Can the effects of antidepressants be observed in the first two weeks of treatment? *Neuropsychopharmacology* 15: 390–394

Ruiz S, Chu P, Sramek J, Rotavu E, Herrera J (1996) Neuroleptic dosing in Asian and Hispanic outpatients with schizophrenia. *Mt Sinai J Med* 63 (5–6): 306–309

Sachse C, Brockmoller J, Bauer S, Roots I (1997) Cytochrome P450 2D6 variants in a Caucasian population: allele frequencies and phenotypic consequences. *Am J Hum Genet* 60 (2): 284–295

Stassen HH, Delini-Stula A, Angst J (1993) Time course of improvement under antidepressant treatment: a survival-analytical approach. *Eur Neuropsychopharmacol* 3: 127–135

Stassen HH, Angst J, Delini-Stula A (1997) Delayed onset of action of antidepressant drugs? Survey of recent results. *Eur Psychiatry* 12: 166–176

Stassen HH, Angst J (1998) Delayed onset of action of antidepressants. Fact or fiction? *CNS Drugs* 9 (3): 177–184

Tollefson GD, Holman SL (1994) How long to onset of antidepressant action: a meta-analysis of patients treated with fluoxetine or placebo. *Int Clin Psychopharmacol* 9: 245–250

Van Praag HM (1998) Inflationary tendencies in judging the yield of depression research. *Neuropsychobiology* 37: 130–141

Ware MR, Emmanuel NP, Jones C (1998) Prevalence and type of herbal product use among college students. NCDEU 38th Annual Meeting, Abstracts 66

Wender PH, Kety SS, Rosenthal D, Schulsinger F, Ortmann J, Lunde I (1986) Psychiatric disorders in the biological and adoptive families of adopted individuals with affective disorders. *Arch Gen Psychiatry* 43: 923–929

Willner P (1989) Sensitization to the actions of antidepressant drugs. In: Emmett-Oglesby MW, Goudie AJ (eds) Psychoactive drugs tolerance and sensitization. Clifton, Humana Press, 407–459

Commentary on the limitation of antidepressants in current use

Stephen M. Stahl

Clinical Neuroscience Research Center, 8899 University Center Lane, San Diego, California 92122, USA; and Department of Psychiatry, University of California San Diego

Introduction

Antidepressants in current use are generally classified by their immediate pharmacological actions [1, 2]. Thus, there are two classical pharmacological classes that include over a dozen individual drugs and the six newer pharmacological classes that include at least 10 specific agents (Tab. 1) [1, 2]. As a group, these antidepressants are important therapeutic agents with many favorable characteristics. The newer antidepressants have a better safety profile than the classical antidepressants. All antidepressants, however, are generally recognized as effective for the majority of patients with major depressive disorder, and also have an ever expanding portfolio of

Table 1. Classes of antidepressants

Pharmacological class	examples
Classical tricyclic antidepressants	clomipramine; imipramine; amitriptyline, nortriptyline, protriptyline, maprotiline, amoxapine, doxepin, desipramine, trimipramine, lofepramine
Classical monoamine oxidase inhibitors	phenelzine, tranylcypromine, isocarboxazid
Serotonin selective reuptake inhibitors	fluoxetine paroxetine fluvoxamine sertraline citalopram
Noradrenergic selective reuptake inhibitors	reboxetine
Noradrenergic and dopaminergic reuptake inhibitors	bupropion
Serotonin 2A antagonist and serotonin reuptake inhibitors	nefazodone trazodone
Dual serotonin and noradrenergic reuptake inhibitors	venlafaxine
Alpha 2 antagonist	mirtazapine

therapeutic actions in other disorders, such as panic disorder, obsessive compulsive disorder, generalized anxiety disorder, social phobia, post-traumatic stress disorder, and many others. Over half of patients with depression treated with any of the currently available antidepressants will respond to treatment, and over 90% will respond if several antidepressants are tried in sequence, or combinations of antidepressants and other drugs are administered [3–5]. Antidepressants also reduce relapse rates for the first 6 to 12 months after improving from an episode of major depressive disorder [6]. So what could possibly be the limitations of drugs with so many positive features?

This chapter will discuss five attributes of the currently available antidepressants that constitute important limitations to their use in treating depression:

(1) slow onset of action;
(2) better usefulness in causing an antidepressant treatment *response* (i.e. 50% or more improvement in symptoms) than *remission* of symptoms (i.e. elimination of symptoms);
(3) only partially able to reduce *relapses* and *recurrent episodes*;
(4) *unable* to cause response or remission at all in many patients;
(5) not ideal for *bipolar* depression, as they do not treat the manic, mixed or rapid cycling dimensions of this illness, and when they treat the depressed phase, can actually induce mixed mania and rapid cycling states.

Slow onset of action

The "holy grail" of antidepressant therapy is the possibility that new agents may have a rapid onset of action, faster than the current agents whose response begins at best within hours to days and usually take up to several weeks to reduce overall symptoms significantly [1, 2]. Complete remission of symptoms often takes many months [3, 4]. Although some observers argue that antidepressants have an *onset* of a noticeable treatment response shortly after administration, and it is only cumulative symptomatic responses that are delayed [7], it is clear that relief of depression is certainly not as rapid as relief of anxiety after administration of a benzodiazepine, or relief of pain after administration of an opiate.

Why can't an antidepressant act as fast and as robustly as a benzodiazepine or an opiate? Given the theories that the mechanism of action of current antidepressants is due to changes in gene expression [1, 8], it is not surprising that the onset of antidepressant effects are dependent upon the time it takes for the rate of synthesis of gene products to be changed, and for these gene products to then alter neuronal functioning (Fig. 1). This may ultimately limit the pace of therapeutic action of any antidepressant agent in the future.

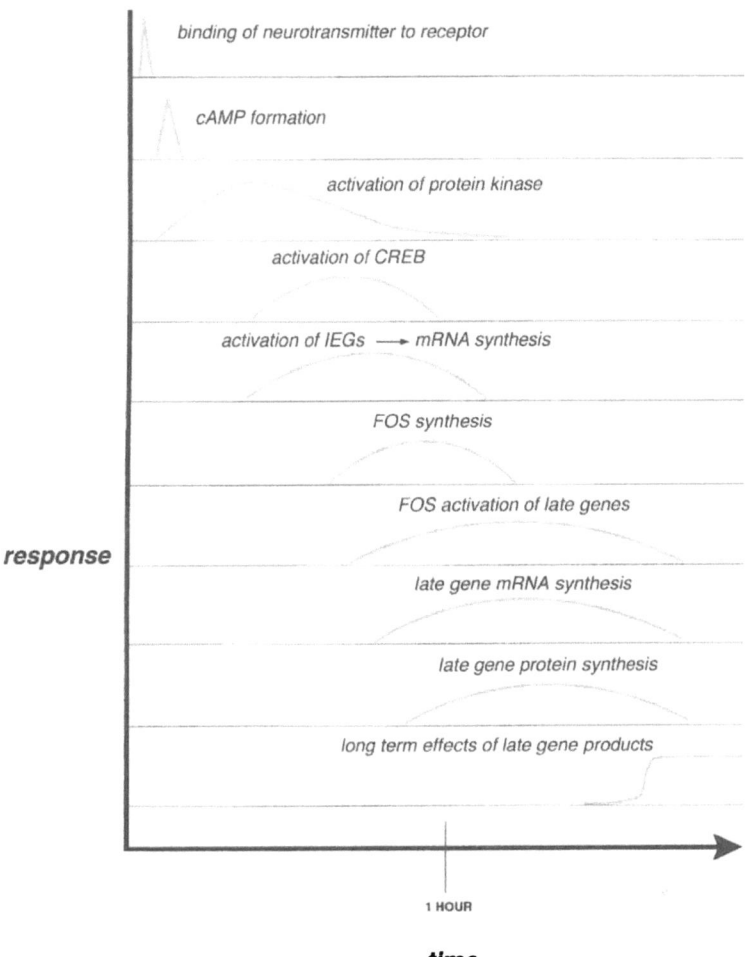

Figure 1. Onset of therapeutic action of an antidepressant may be delayed if receptor-mediated activation of gene expression is required. Thus, antidepressant actions may be dependent upon the ultimate physiological actions of late gene products (bottom). This may be delayed for days to weeks following activation of monoamine receptors (top).

For example, the mechanism of action of serotonin selective reuptake inhibitors seems to be linked to the desensitization of certain neurotransmitter receptors, particularly the 5HT1A somatodendritic autoreceptors [1]. If this desensitization is mediated by decreasing gene expression of these receptors, the onset of enhanced release of serotonin, and presumably an antidepressant action, may have to await an adjustment in the synthesis of these receptors. This could take days to weeks. Furthermore, if increases in the synthesis of neurotrophic factors in the brain (such as brain derived neurotrophic factor, or BDNF) prove to be necessary for full antidepressant

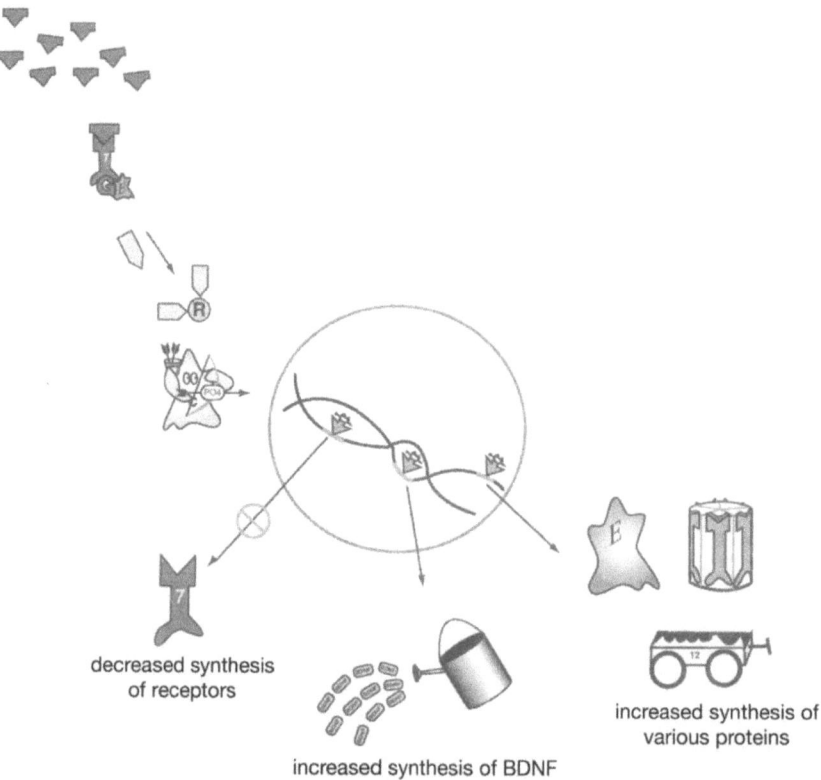

Figure 2. The monoamine hypothesis of antidepressant action on gene expression proposes that antidepressants exert their therapeutic actions *via* changes in the expression of critical genes in the appropriate neuronal sites. Thus, decreases in the synthesis of neurotransmitter receptors as well as increases in the synthesis of various key gene products, such as neurotrophic factors may be necessary for full antidepressant effects to become manifest. One candidate for an important neurotrophic factor in brain is BDNF or brain derived neurotrophic factor.

actions, as suggested by a prominent contemporary theory of depression (Fig. 2) [8], the actions of newly synthesized neurotrophic factors will take considerable time to develop, probably several weeks to months.

Short circuiting antidepressant actions mediated by changes in synthesis of gene products may require a drug to directly activate or inactivate a gene, or to substitute for the gene product directly. Some hint that new therapeutics may be able to bypass a time-consuming activation of genes is suggested by the treatment responses of women whose depressions are responsive to estrogen [9]. These antidepressant treatment responses may be relatively rapid perhaps due the action of estrogen directly upon estrogen response elements of the genome (Fig. 3). In order for future antidepressants to be very rapidly acting, it will require better understanding of which gene products are critical for the antidepressant response, so that they can be mimicked immediately by future drugs.

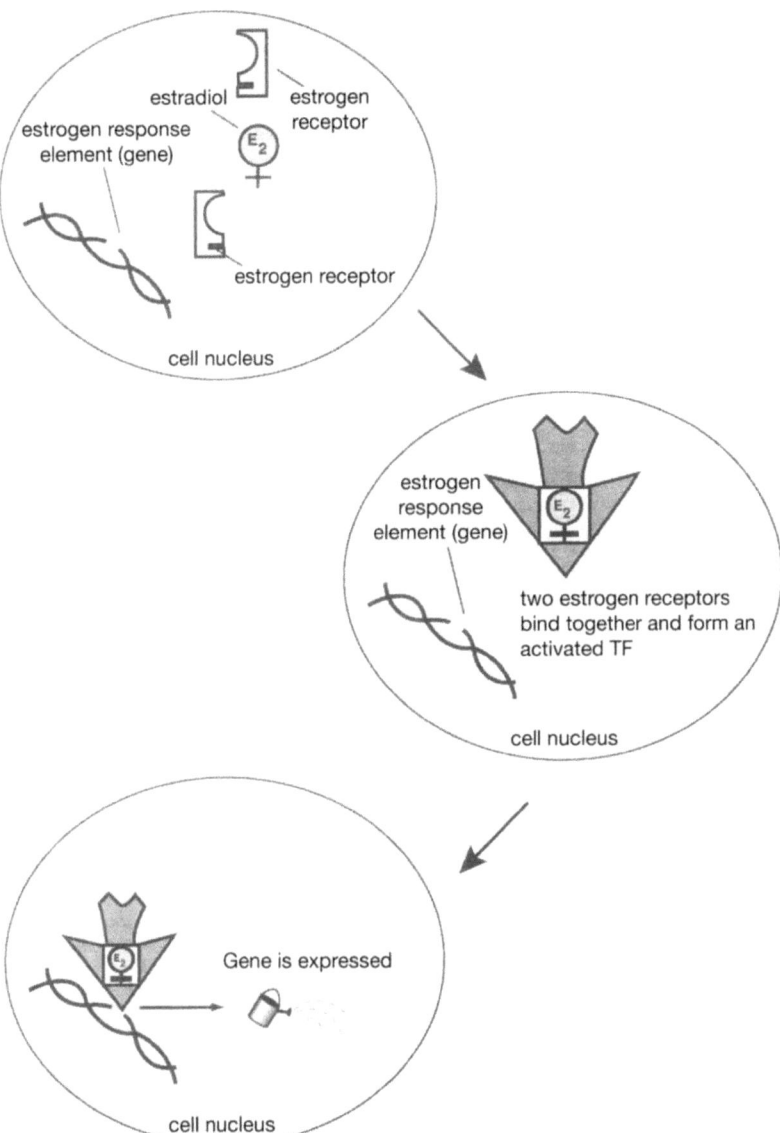

Figure 3. Estrogens activate gene expression by binding to their receptors in the cell nucleus, thus forming a transcription factor that can activate estrogen response elements in various neurons. Since this process bypasses the neurotransmitter receptor-induced cascade shown in Figures 1 and 2, it may be faster in onset.

Antidepressant onset of action for current drugs is particularly slow if one is targeting complete remission of symptoms, which usually takes months of treatment. This will be explored in the following section.

Better for response than for remission and recovery

Classically, antidepressant efficacy has been evaluated in studies where the endpoint is a *treatment response* defined as a 50% or greater reduction of symptoms on the Hamilton depression rating scale or other comparable rating instruments (Fig. 4) [1, 3, 4, 6]. Nearly all studies of antidepressants show that treatment responses are roughly comparable for any two antidepressants, namely about 40 to 60% of depressed patients will respond after 2 months of treatment [1, 3, 4, 10]. Furthermore, up to 90% of patients will respond to one or another antidepressant or combination of antidepressants if enough are tried in sequence [5].

Focussing on treatment response tends to make the efficacy of current antidepressant treatments look fairly good, but can obscure the limitations of current antidepressants when a more clinically relevant outcome, such as return to normal functioning, is desired. Although a 50% improvement of symptoms may have been an adequate endpoint for regulatory agencies assessing current antidepressant efficacy prior to marketing, it is not an appropriate endpoint for the current and future treatments for depression. Patients with less than a complete reduction of symptoms have an increased risk of relapse, increased risk of suicide and still have disabling residual

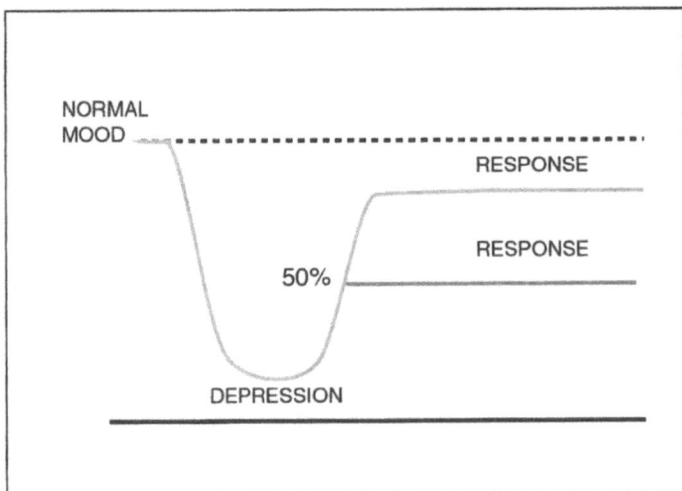

Figure 4. A response to an antidepressant is defined as 50% or more improvement in baseline symptoms.

symptoms compared to patients who are in full remission [3, 4, 11–16]. Since it is possible for current antidepressants to eliminate depressive symptoms at least in some patients, it is important to use this as the endpoint for the use of current antidepressants. Unfortunately, perhaps only 20% of patients remit as early as 2 months after antidepressant treatment, with the majority taking much longer to remit, even if they remain compliant with long-term treatment. This lack of robust recovery soon after antidepressant treatment is started is a major limitation of current antidepressants.

More recently, therefore, treatment studies have begun to target *remission* (i.e. elimination of symptoms, defined as scores as low as 7 on the Hamilton depression rating scale for a few months) or *recovery* (i.e. elimination of symptoms for many months) (Fig. 5A). Some studies of current antidepressants suggest that remission rates of 20% after 2 months of treatment may rise to as high as 50% after 6 months, perhaps to as high as 75% or more after 2 years of treatment [12–16]. Future antidepressants must target remission, not response, and attempt to increase remission rates so that the majority of patients do not have to wait for 6 to 24 months for a remission as they must now.

Hints that new antidepressants might possibly increase remission rates come from studies of some of the newer antidepressants with dual pharmacological mechanisms of action. That is, antidepressants that act by boosting both serotonergic and noradrenergic neurotransmission seem to have better (or at least faster onset) remission rates than antidepressants acting

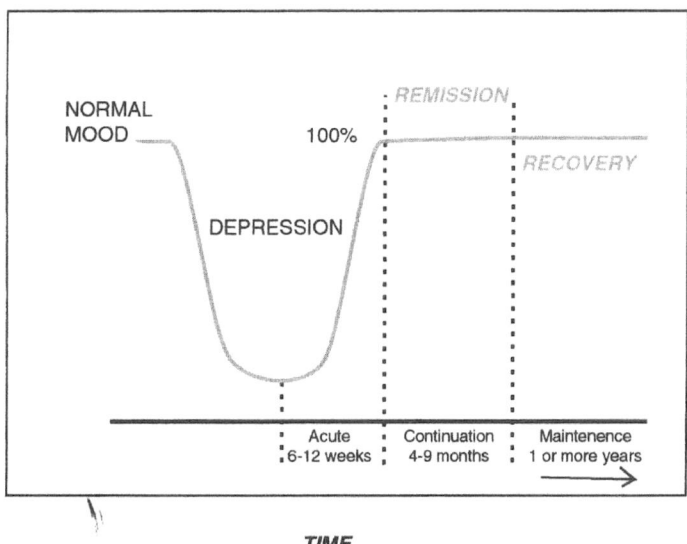

Figure 5A. Remission from depression is defined as complete reduction of depressive symptoms, or a return to a normal state. If remission lasts for several months, it is called recovery.

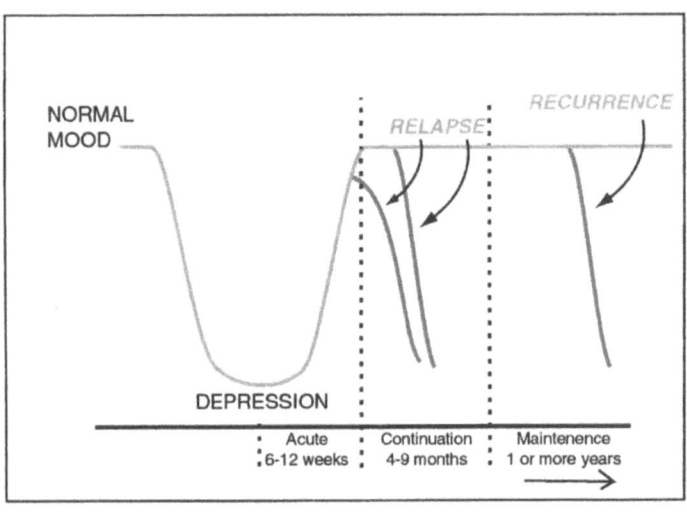

Figure 5B. Relapses into another episode of depression can occur either before there is a full remission, or within the first few months following remission. However, if depression returns after several months of a remission, it is considered to be a recurrence.

by a serotonin selective mechanism [17, 18]. Thus, the dual serotonin and norepinephrine reuptake inhibitors venlafaxine and several tricyclic antidepressants may have higher remission rates after 2 months of treatment than the serotonin selective reuptake inhibitors. Dual-acting mirtazapine also may have higher remission rates, as does a combination of one SSRI and one noradrenergic reuptake inhibitor. Since these "dual action" antidepressant strategies seem to increase remission rates in depressed patients, they suggest that strategies which synergize among neurotransmitter systems may be more effective than strategies that target a single neurotransmitter system [17]. Perhaps the future will bring new neurotransmitter systems, such as substance P, to target in addition to the currently targeted monoamine systems of serotonin, norepinephrine and dopamine. Adding such a system to the monoamines, rather than targeting a new system selectively may prove to bring synergy to antidepressant actions, and thus boost remission rates, especially early in treatment.

Only partially able to reduce relapses and recurrent episodes

Unfortunately, attaining a treatment response or even full remission from depression does not guarantee long-term recovery if current antidepressant treatments are continued [3, 4, 6, 11–16]. Although antidepressant responders experience less of a chance of relapse into another episode of

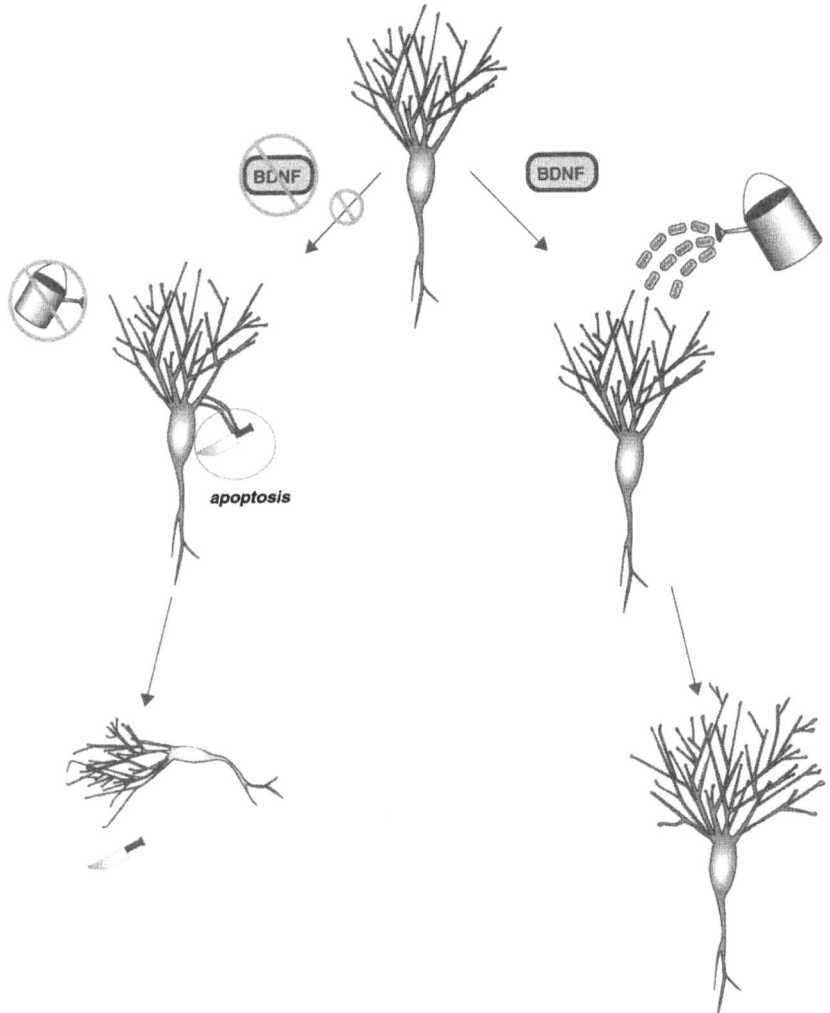

Figure 6. Normally, the brain makes adequate amounts of neurotrophic factors such as brain derived neurotrophic factors (BDNF). When this happens, the neurons are "fertilized" and sprout many dendritic branches (shown on the right). However, under stress, it is hypothetically possible that BDNF synthesis is turned off. This could lead to death of neurons, perhaps by a process of neuronal suicide known as apoptosis. This degenerative process could underlie the increasing risk of relapses and recurrences in patients with many episodes of depression, and the development of resistance to treatment with antidepressants.

depression if they continue their medications for another 6 to 12 months, there is still significant risk of relapse on drug treatment (sometimes called "poop-out") (Fig. 5B). Thus, an antidepressant responder has a 50% chance of relapsing into another episode of depression in the first 6 to 12 months after stopping the antidepressant, but only a 10–20% chance of relapsing if antidepressants are continued during this time. Current antidepressants thus reduce relapse in responders, but do not eliminate relapses, even when antidepressants are continued chronically. New antidepressants must reduce relapse rates to less than 10–20% during treatment for the first 6 to 12 months.

Even more problematic is the fact that those patients who are fortunate enough to go on to experience a long-term recovery from depression still have a very high rate of recurrence of another episode of illness (Fig. 5) [6, 11–16]. It is not clear that current antidepressants reduce this long-term recurrence rate if antidepressants are continued for several years, but it is very important for future antidepressants to show this action. Not only is recurrence itself an obvious form of suffering, but there is disturbing evidence that depressive episodes themselves may have long-term damaging actions upon the brain. That is, neuroimaging studies showing atrophy of frontal cortex and hippocampus support the hypothesis that neurotrophic factors may be deficient in depression (Fig. 3) [1, 8]. Since the more depressive episodes that one has, the more depressive episodes one is destined to get in the future, especially if there has been incomplete recovery between episodes [6, 11–16], there is a suggestion that a slow neurodegenerative process in the brain may underlie chronic and recurrent depression (Fig. 6) [1, 8]. This may also lead to resistance to treatment with antidepressants. Thus, future antidepressants must reduce recurrent episodes of depression to prevent a potentially progressive course to this illness.

Unable to cause any response or remission in many patients

Not all patients with depression remit on antidepressants, nor even respond to antidepressants. These so-called non-responders, treatment resistant, or treatment refractory patients are an increasingly troublesome population of depressed patients, especially in a modern psychiatrist's practice. Up to 30% of patients with depression have a poor outcome with antidepressant treatment, including not only nonremitters and nonresponders, but also those with chronic, recurrent depressive episodes who have lost their responsiveness to antidepressants [5, 11–16]. Currently, these cases are treated with the varying antidepressant monotherapies with differing mechanisms of action (Tab. 1), and when these fail, by combinations of antidepressants and various other agents. New treatments should target such cases, as they represent a large percentage of health care utilization and disability among the population of depressed patients. Perhaps targeting neurotransmitter systems other than monoamines will be necessary for such patients who fail to respond to any agent that currently targets the monoamine systems.

Limited efficacy in bipolar disorder

Current antidepressants can reduce depressive symptoms in bipolar patients, particularly in the presence of a mood stabilizer. However, antidepressants are not only unable to treat the manic phase of this illness, they often decompensate bipolar patients and induce manic episodes when these patients are treated in the depressive phase of their illness with antidepressants. Even in the presence of mood stabilizers and antipsychotic drugs, current antidepressants have the propensity to destabilize bipolar patients by inducing mixed states of agitated, mixed mania and depression, rapid cycling between mania and depression, or both. Future antidepressants should be able to treat these various dimensions of depression in bipolar patients without decompensating them, and perhaps with the additional ability to treat the manic elements of the illness so that current mood stabilizers and/or antipsychotics may possibly be avoided.

Summary and conclusions

Antidepressants have revolutionized the treatment not only of affective disorders, but anxiety disorders as well. The classical tricyclic antidepressants and monoamine oxidase inhibitors are clearly effective, and the numerous newer classes of antidepressants clearly safe and effective in the treatment of depression. Thus, antidepressants are important therapeutic agents in the armentarium of psychopharmacology. On the other hand, numerous distinct limitations exist for these agents which future antidepressants will hopefully surmount. Five such limitations are discussed in this chapter. Firstly, current antidepressants have a slow onset of therapeutic action, taking about 2 months for a therapeutic response and up to 24 months to cause symptomatic remission. Secondly, current antidepressants are much better in causing an antidepressant treatment *response* (i.e. 50% or more improvement in symptoms) than *remission* of symptoms (i.e. elimination of symptoms). Since a return to full function and "normality" are the real goals of antidepressant treatment, this is indeed a serious limitation for those patients who fail to remit. Thirdly, contemporary antidepressants are only partially able to reduce relapses and recurrent episodes, even if treatment is continued. This can lead to unfavorable long-term outcomes for patients with many episodes of depression. Fourthly, current antidepressants are unable to cause response or remission at all in many patients. These treatment resistant patients are particularly disabled by their illness and are a major treatment issue for most psychiatrists. Finally, current antidepressants are far from ideal for bipolar depression, as they do not treat the manic, mixed or rapid cycling dimensions of this illness, and when they treat the depressed phase, can actually induce mixed mania and rapid cycling states.

Future antidepressants that address these limitations will improve significantly the already impressive usefulness of the current armamentarium of antidepressant drugs.

Acknowledgement
Figures reproduced with permission from Stahl SM, Essential Psychopharmacology, 2nd edition, Cambridge University Press, New York.

References

1. Stahl SM (2000) *Essential psychopharmacology*, 2nd edition, Cambridge University Press, New York
2. Leonard BE (1997) *Fundamentals of psychopharmacology*. John Wiley and Sons Ltd, Chichester
3. *Depression in primary care, volume 2, treatment of major depression*; clinical practice guideline, number 5 (April 1993) U.S. Department of Health and Human Services, AHCPR Publication No. 93-0551, Public Health Service, Agency for Health Care Policy and Research, Rockville, MD
4. *Depression in primary care, volume 1, detection and diagnosis*; clinical practice guideline, number 5, (April 1993) U.S. Department of Health and Human Services, AHCPR Publication No. 93-0551, Public Health Service, Agency for Health Care Policy and Research, Rockville, MD
5. Thase ME, Rush AJ (1995) Treatment resistant depression. In: FE Bloom, DJ Kupfer DJ (eds): *Psychopharmacology: the fourth generation of progress*. Raven Press, New York
6. Kupfer DJ, Frank E, Perel JM et al (1992) Five year outcome for maintenance therapies in recurrent depression. *Arch Gen Psychiat* 49: 769–773
7. Duman RF, Heninger CR, Nestler EJ (1997) A molecular and cellular theory of depression. *Arch Gen Psychiatry* Vol. 54: 597–606
8. Stahl SM (1998) Basic Psychopharmacology of Antidepressants, part 2: estrogen as an adjunct to antidepressant treatment. *J Clin Psychiat* 59 (suppl 4): 15–24
9. Prien RF, Robinson DS (eds) (1994) *Clinical evaluation of psychotropic drugs: principles and guidelines*. Raven Press, New York
10. Robins LN, Regier DA (1991) *Psychiatric disorders in America: the epidemiologic catchment area study*. New York, The FreePress (Macmillan Inc.)
11. Keller MB, Shapiro RW, Lavori PW et al (1992) Recovery in major depressive disorder: analysis with the life table and regression models. *Arch Gen Psychiatry* 39: 905–910
12. Keller MB, Shapiro RW, Lavori PW et al (1992) Relapse in major depressive disorder: analysis with the life table. *Arch Gen Psychiatry* 39: 911–915
13. Keller MB, Klerman GL, Lavori PW et al (1984) Long term outcome of episodes of major depression: clinical and public health significance. *JAMA* 252: 788–792
14. Keller MB, Lavori PW, Mueller TI et al (1992) Time to recovery, chronicity and levels of psychopathology in major depression: a 5 year prospective follow up of 431 subjects. *Arch Gen Psychiatry* 49: 809–816
15. Keller MB, Lavori PW, Rice J et al (1986) The persistent risk of chronicity in recurrent episodes of nonbipolar major depressive disorder: a prospective follow-up. *Am J Psychiatry* 143: 24–28
16. Steffens DC, Krishnan KRR, Helms MJ (1997) Are SSRIs better than TCAs? Comparison of SSRIs and TCAs: A meta-analysis. *Depression and Anxiety* 6: 10–18
17. Anderson IM (1998) SSRIs vs. tricyclic antidepressants in depressed inpatients: a meta-analysis of efficacy and tolerability. *Depression and Anxiety* 7: 11–17
18. Einarson TR, Arikian SR, Casciano J et al (1999) Comparison of extended release venlafaxine, selective serotonin reuptake inhibitors and tricyclic antidepressants in the treatment of depression: a meta analysis of randomized controlled trials. *Clin Therapeutics* 21: 296–308

19 Stahl SM (1997) Are two antidepressant mechanisms better than one? *J Clin Psychiatry* 58 (8): 339–341
20 Nelson JC, Mazure CM, Bowers MB et al (1991) A preliminary, open study of the combination of fluoxetine and desipramine for rapid treatment of major depression. *Arch Gen Psychiat* 48: 303–307
21 Gelenberg AJ, Bassuk EL (1997) *The practitioners guide to psychoactive drugs*, 4th edition. Plenum Medical Book Company, New York

Mechanisms of action

Mechanisms underlying the speed of onset of antidepressant response

Caroline McGrath and Trevor R. Norman

Dept. Psychiatry, University of Melbourne, Australia

Introduction

Depression is an incapacitating disorder with serious health and economic consequences. It has an estimated cost of more than $43 billion annually in the US (Greenberg et al., 1993) and by the year 2020 it will be the second leading cause of disease burden behind ischaemic heart disease (The Global Burden of Disease and Injury Series, 1996). Despite extensive research, very little is known about the aetiology of depressive illness or the mechanism of action of drugs used in its treatment. Although effective, there are a number of disadvantages to the use of antidepressants. Firstly, there is a delay of 3–4 weeks before any therapeutic effects of these drugs are seen. This is a serious problem since there is an increased incidence of suicide in depressed patients and a lifetime incidence of suicide of 15% has been reported from follow-up studies (Smith and Weissman, 1992; Norman and Leonard, 1994). Secondly, antidepressants, particularly the early generation, are often associated with unpleasant side effects such as dry mouth, blurred vision, urinary hesitancy, dizziness on standing, which reduce patient compliance. Improvements have been made with newer antidepressants which are based on the same mechanism of action. Venlafaxine blocks the reuptake of noradrenaline and serotonin but lacks the alpha 1, cholinergic and histaminergic receptor blocking properties of the TCA's, while nefazodone, the 5-HT_2 receptor antagonist, has less side-effects than the SSRI's. Unfortunately, like previous antidepressants, these newer agents exert their effects either on the serotonergic system, the noradrenergic system or both and are therefore unlikely to offer any advantages over earlier drugs in terms of speed of onset (Broekkamp et al., 1995). In addition, approximately 30% of patients treated with a therapeutic dose of an antidepressant fail to respond after several weeks of treatment. There is a need therefore for the development of other novel antidepressants with a more rapid onset of action and a reduced side effect profile than those currently available.

Can an early onset of antidepressant action be detected?

According to Montgomery, 1992, the onset of antidepressant activity is defined as "the time at which statistical separation is first seen between an antidepressant and a comparator, either active or placebo, and at which there is concomitant, clinically relevant decrease on the scale for the user" (cited in Derivan, 1995). There is evidence to suggest that this is possible soon after treatment. A rapid antidepressant response has been reported following electroconvulsive therapy, with improvement noted within 1 week. In addition, an antidepressant response has been found following one night of sleep deprivation (Blier and Bergeron, 1997). These, and similar findings, coupled with the evidence from preclinical studies of rapid receptor adaptation, suggests that a fast onset of antidepressant action is attainable.

Measurement of antidepressant effects

Various rating scales have been employed in the literature for the determination of antidepressant effect. Of these the most widely used scales are the Hamilton Depression Rating Scale (HAM-D), the Montgomery Åsberg Depression Rating Scale (MADRS) and the Clinical Global Impressions Scale (Derivan, 1995), of which the HAM-D is the most widely used in drug trials in English speaking countries (Kellner, 1994). The HAM-D has been criticised on the basis that an improvement in the HAM-D score may not necessarily reflect an antidepressant response (Prien and Levine, 1984), possibly due to the fact that the HAM-D measures insomnia and somatic symptoms as well as disturbances of mood, so the overall score may not be a true reflection of the individuals depressive state (Müller and Möller, 1998). In addition, if meaningful information regarding the onset of action of antidepressant drugs is to be obtained, then it is essential that rating scales which are sensitive to changes in severity of depression are used. The HAM-D has been criticised by Montgomery and Asberg (1979) for being insensitive to change during treatment. The MADRS, on the other hand, is designed to be particularly sensitive to change and may give shorter improvement delays than the HAM-D (Müller and Möller, 1998). Unfortunately the MADRS has not been as widely used as the HAM-D, so its utility for detecting an early onset of antidepressant action has not been well evaluated (Norman and Leonard, 1994). Maier et al. (1988) reviewed the ability of the HAM-D, the MADRS and the Bech-Rafaelsen Melancholia Scale (BRMS) to detect change in severity of depression following antidepressant treatment. They concluded that the BRMS is superior to both the HAM-D and the MADRS with respect to sensitivity to change and suggested that this study questions the usefulness of the HAM-D alone when testing the efficacy of antidepressants.

Prien and Levine (1984) outlined other methodological problems with determining early onset of antidepressant action. Firstly it is necessary that clinically meaningful changes are clearly defined. Secondly, that clinical assessments are conducted at regular intervals, to ensure that any therapeutic response is detected. Furthermore, it is essential that adequate doses of both the new drug and a reference drug are administered. Unfortunately, very often patients can not be commenced on optimal doses of the drugs due to adverse side-effects (Norman and Leonard, 1994), thus limiting the ability of clinical trials to detect an early response.

Targets for antidepressant action

Early theories on the mechanism of action of antidepressants focused on the role of the biogenic amines in depression. The "monoamine" hypothesis stated that depression was the result of a decrease in biogenic amines, namely noradrenaline and serotonin. Problems with this were soon evident and it is now generally believed that neuroadaptive receptor changes are responsible for the therapeutic effect of antidepressants.

Noradrenergic receptor changes

Much of the early work on the mechanism of action of antidepressant drugs focused on changes in the noradrenergic receptor system. Chronic antidepressant treatment results in a decrease in noradrenaline (NA) or isoproterenol-stimulated cyclic AMP (cAMP) production in rodent brain slices, an effect which is usually accompanied by a decrease in the number of β-adrenoceptor binding sites (Vetulani et al., 1976; Wolfe et al., 1978; Sulser et al., 1978). This down regulation is believed to be due to increased NA available at the synaptic cleft (Ordway et al., 1988), as a consequence of down regulation of presynaptic α_2 autoreceptors. Enhanced NA release has been shown to increase the rate at which antidepressant administration modifies β-adrenoceptor activity (Crews et al., 1981; Kendall et al., 1982). Strategies aimed at increasing the speed of onset of these neuroadaptive changes have been employed, in an attempt to produce an earlier onset of antidepressant action. One mechanism that has been suggested to enhance this downregulation is by inhibition of presynaptic α_2-adrenoceptors. A combination of desipramine with the α_2-adrenoceptor antagonist phenoxybenzamine has been reported to produce a more rapid decrease in the number of β-adrenoceptor binding sites than desipramine alone (Crews et al., 1981; Scott and Crews, 1983). Maximum neuroadaptive changes were evident after 3 days of treatment with the combination as opposed to 7 days of treatment with desipramine alone. Similar findings have been reported with the use of yohimbine (Wiech and Ursillo, 1980). However, Charney et al. (1986) failed to find any beneficial effects of combined treatment with yohimbine and desipramine in patients with treatment resistant depression.

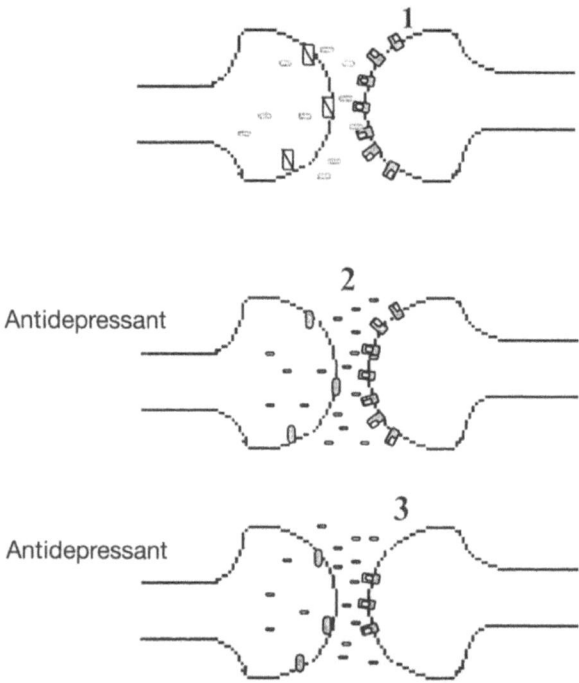

Figure 1. Summary of events at the noradrenergic receptor following antidepressant treatment. *1* Post-synaptic receptor upregulated (due to decreased availability of NA). *2* Increased NA release, due to administration of TCA. *3* Down-regulation of post-synaptic receptors

However, it is likely that the use of this group of patients would reduce the chances of observing an early onset of antidepressant action (Norman and Leonard, 1994). Sachs and colleagues (1986) reported a dramatic response in three patients (meeting DSM III criteria for major depression) with the use of Electro Convulsive Therapy (ECT) in combination with yohimbine. Although the results appear to be promising, clearly further placebo controlled clinical studies are needed.

Manipulation of the serotonergic system also results in desensitisation of β-adrenoceptors. Baron et al., (1988) reported a significant decrease in β-adrenoceptor number in the rat following 4 days of treatment with a combination of fluoxetine and desipramine. Down-regulation of β-adrenoceptor has also been reported following treatment with selective serotonin reuptake inhibitors (SSRI's) alone (Dennis et al., 1994; McGrath, 1996) although this has not been a consistent finding (Mishra and Sulser, 1978; Garcha et al., 1985; Nelson, 1990).

The question remains whether β_1-adrenoceptor downregulation is a necessary or sufficient condition for an antidepressant effect. Theoretically, at least, the development of a specific β_1 agonist with selectivity for the central nervous system could provide an answer to this question. In this context the development of agents such as clenbuterol, a selective β_1 agonist, have not shown promise in the clinic as antidepressants. Perhaps β_1-adrenoceptor downregulation is not a necessary condition for an antidepressant effect. The lack of effect of citalopram (Garcha et al., 1985) and fluoxetine (Mishra and Sulser, 1978) on β_1-adrenoceptors in rodents, despite their proven clinical effectiveness, further supports this notion.

Serotonergic receptor changes

Over the last 20 years our knowledge of the serotonergic system has increased dramatically, and with it, there has been a growing interest in the importance of this system in the aetiology of depressive disorders. To date, seven different subtypes of serotonergic receptors have been identified and are generally divided into $5\text{-}HT_1$, $5\text{-}HT_2$, $5\text{-}HT_3$, $5\text{-}HT_4$, $5\text{-}HT_5$, $5\text{-}HT_6$ and $5\text{-}HT_7$ receptors, many of which have been further subdivided. The $5\text{-}HT_1$ receptors for example have been subdivided into $5\text{-}HT_{1A}$, $5\text{-}HT_{1B}$, $5\text{-}HT_{1D}$ and $5\text{-}HT_{1E}$ (Frances and Khidichian, 1990), although evidence suggests that the $5\text{-}HT_{1B}$ and the $5\text{-}HT_{1D}$ receptors may be species homologues of the same gene (Martin and Puech, 1991). Recently there has been a growing interest in the role of the $5\text{-}HT_1$ receptors in the treatment of depressive illness. $5\text{-}HT_{1A}$ and $5\text{-}HT_{1B}$ receptors function as somatodendritic and terminal autoreceptors respectively, stimulation of which has an inhibitory effect on 5-HT release. Chronic antidepressant treatment has been reported to result in increased 5-HT neurotransmission, an effect believed to be responsible for the therapeutic effect of the selective serotonin re-uptake inhibitors. It has been suggested therefore that the delay in the onset of action of antidepressants may be due to an adaptation of the $5HT_{1A}$ somatodendritic and $5\text{-}HT_{1B/1D}$ terminal autoreceptors (Blier and De Montigny, 1994).

In effect, the initial rise in extracellular 5HT following antidepressant administration feeds back on $5HT_{1A}$ and $5\text{-}HT_{1B/1D}$ autoreceptors in somatodendritic sites and leads to an initial reduction of firing and synthesis of 5HT through a negative feedback loop. With repeated administration of antidepressants, a desensitisation of this feedback loop occurs as the autoreceptors adapt and downregulate. This allows the neuron to fire more frequently and begin again the synthesis and release of 5HT, but this time there is reuptake blockade at the nerve terminals leading to an overall increase of 5HT at postsynaptic sites. Indeed, electrophysiological studies in rats have shown that repeated administration of SSRI's and various other antidepressants, leads to a functional desensitisation of somatodendritic $5HT_{1A}$ and $5\text{-}HT_{1B}$ receptors (Blier and De Montigny, 1994). The timing of this desensitisation correlates well with the delay in onset of action of antidepressants. Theoretically then, if the observed delay is a function of $5\text{-}HT_{1A}$ and $5\text{-}HT_{1B}$ receptor desensitisation, then selective blockade of

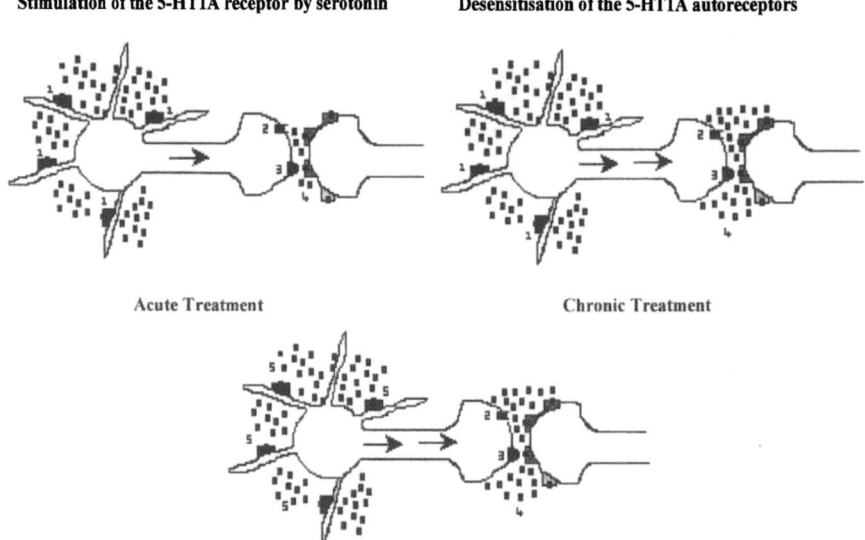

Blockade of 5-HT1A receptor with pindolol

1 = 5-HT$_{1A}$ receptor; 2 = 5-HT$_{1B}$ receptor; 3 = 5-HT Transporter + SSRI; 4 = 5-HT, 5 = 5-HT$_{1A}$ receptor + Pindolol

Figure 2. Summary of the events at the serotonergic receptor following treatment with an SSRI alone and in combination with the 5-HT$_{1A}$ receptor antagonist pindolol.

these receptors (which would mimic this desensitisation), should produce a more rapid antidepressant response. Currently, there are no 5-HT$_{1A}$ receptor antagonists approved for use in humans, however one combination that has been reported to be effective in reducing the delay in onset of antidepressant action is the use of the β-blocker/5-HT$_{1A}$ receptor antagonist pindolol in combination with an SSRI. This combination has been shown in some controlled studies to speed the onset of action of various antidepressant drugs (Blier and Bergeron, 1995; Blier and De Montigny, 1994; Bordet et al., 1998) although the results have not always been promising, with some studies failing to find any beneficial effects of combining pindolol with an SSRI (Dinan and Scott, 1996; Berman et al., 1997).

Although very little is known regarding the clinical importance of the 5-HT$_{1B}$/5-HT$_{1D}$ receptors (Pauwels, 1997), there is a vast body of preclinical evidence suggesting an important role for these receptors in depressive illness. Firstly, 5-HT$_{1B}$ receptors have been reported to be increased in the cortex, hippocampus and septum of rats in the Learned Helplessness Model (Edwards et al., 1991). Secondly, both 5-HT$_{1B}$ autoreceptors and 5-HT$_{1B}$ heteroreceptors are desensitised following chronic antidepressant treatment (Blier et al., 1988; Bolanos-Jimenez et al., 1994;

Moret and Briley, 1990). Thirdly, administration of the 5-HT$_{1B}$ agonist 1-(3-Trifluoromethylphenyl)piperazine (TFMPP), antagonised the antidepressant effects of fluvoxamine and citalopram in the Learned Helplessness Model (Martin and Puech, 1991) and desipramine in the Forced Swim Test (Cervo et al., 1989), possibly due to its negative actions on 5-HT neurotransmission (Martin and Puech, 1991). If this is the case then selective blockade of these receptors should increase neurotransmission and in doing so, may offer a potential novel antidepressant therapy (Briley and Moret, 1993). Roberts et al., 1997 reported that antagonism of the 5-HT$_{1D}$ receptors in the guinea pig, rather than increasing 5-HT levels as expected, resulted in a significant decrease in 5-HT concentrations. Although the reason for this was unclear it has been suggested that it may be due to the fact that antagonism of the 5-HT$_{1D}$ receptors results in increased extracellular 5-HT, which would stimulate the 5-HT$_{1A}$ receptors, thus decreasing 5-HT cell firing and release. If this is the case, then blockade of both the 5-HT$_{1A}$ and 5-HT$_{1B}$ autoreceptors may be a more powerful way to potentiate the effects of an SSRI than blockade of either receptor alone (Sharp et al., 1997). Sharp and colleagues (1997) reported a greater increase in 5-HT concentration in the frontal cortex of rats treated with paroxetine and the 5-HT$_{1B}$ receptor antagonist, GR 127935 and the 5-HT$_{1A}$ receptor antagonist, WAY 100635 than with either antagonist alone. This study suggests that an SSRI may have a better therapeutic effect when both receptors are blocked. However, whether blockade of the 5-HT$_{1B}$ receptor alone will lead to an early response in patients remains to be seen and requires further investigation.

Dopaminergic system

The importance of the dopaminergic system as a target for antidepressant action has received relatively little attention due to the belief that noradrenaline and serotonin are the primary receptor systems involved in the pathophysiology of depression. It is noteworthy however that much of the evidence supporting a role for noradrenaline in depression also applies to the dopaminergic system – reserpine depletes stores of dopamine, monoamine oxidase inhibitors (MAOI's) prevent its breakdown, neurotransmitter abnormalities (such as decreased cerebrospinal homovanillic acid (HVA) concentrations in depressed patients) have been reported and there is evidence supporting a therapeutic action of amphetamine and L-dopa (Willner, 1997). In addition, it has been suggested that all currently available antidepressant drugs, irrespective of their acute mechanism of action, have effects on the mesolimbic dopamine system (Ollat, 1988) since chronic antidepressant treatment increases the responsiveness of dopamine D2/D3 receptors (Gessa, 1996; Willner, 1997). Although the mechanism responsible for this effect still remains elusive, it may reflect indirect actions of the drugs on the dopamine system, via projections from the amygdala and hippocampus to the nucleus accumbens (Willner, 1997). The time required for this adaptation to occur may account for the delay in the onset of antidepressant action (Willner, 1997). If adaptation of the mesolimbic dopamine D2/D3 receptors are important for the treatment of depressive illness, then drugs with a direct action on this system may have an earlier onset of antidepressant action than that available with conventional antidepressant drugs.

Support of this hypothesis comes from findings using amineptine. Amineptine, is a potent inhibitor of dopamine reuptake, (although its effects are selective for the mesolimbocortical system (Deniker et al., 1982)), which appears to be devoid of histaminergic or cholinergic properties (Setrakian, 1982). A rapid onset of antidepressant action has been reported with amineptine, with improvement noted on HAM-D scores as early as 7 days of treatment (Kemali, 1989). In a double-blind, controlled study of 51 patients with moderate to severe depression, Van Amerongen (1979) found that amineptine, 200 mg per day, produced a more rapid response than amitriptyline on items such as depressed mood and psychomotor retardation and appeared to be better tolerated. Deniker et al. (1982), in a multicentre study of 1354 patients, reported a rapid antidepressant response with a 34% reduction in HAMD scores evident at 7 days of amineptine treatment. Similar findings have been reported by others (Bornstein, 1979; Paes de Sousa and Tropa, 1989; Dalery et al., 1992; Dalery et al., 1997).

While there is evidence of antidepressant effects of dopamine reuptake inhibition, there is also evidence to suggest antidepressant activity with some D2/D3 receptor antagonists, such as the antipsychotic, sulpiride. In addition to its antidepressant properties, sulpiride may also have an early onset of action. In a double-blind comparison study of L-sulpiride and amitriptyline in bipolar patients, Bocchetta et al. (1993) reported a faster onset of antidepressant action in the L-sulpiride group, with a significant improvement in both anxiety somatisation and specific depressive items such as depressed mood, feelings of guilt, work and activities and retardation following 1 week of treatment. One possible explanation for the antidepressant action, is that at low doses it has been reported to preferentially block dopamine D2/D3 presynaptic autoreceptors (Willner, 1997), thus having the net effect of increasing dopamine neurotransmission.

Evidence suggests a possible role for the dopamine system in the treatment and aetiology of depressive illness. The dopaminergic hypothesis of depression offers an explanation for the antidepressant effects of drugs such as sulpiride and amineptine and identifies possible targets for the development of novel antidepressant drugs with potential early onset of action.

Evidence for an early onset of action – mixed systems

Venlafaxine

Venlafaxine is a noradrenaline and serotonin reuptake inhibitor with little affinity for cholinergic, adrenergic and histaminergic receptor systems. Venlafaxine has been reported to have a more rapid onset of action than the currently available antidepressant compounds with therapeutic response evident within the first 2 weeks of treatment (Derivan et al., 1995; Benkert et al., 1996; Entsuah et al., 1998). The speed of onset of action of venlafaxine compared to placebo was evaluated in 93 patients with DSM-III-R major depression and melancholia using double-blind methodology (Guelfi et al., 1995). After 4 days significant clinical improvement was noted in the venlafaxine treated group on the MADRS score after 1 week on the HAM-D score. This increase in the speed of onset of venlafaxine may be due to its combined noradrenergic and serotonergic reuptake inhibition (Burnett and Dinan, 1998) since a more rapid antidepressant response has been reported following combined treatment of desipramine and fluoxetine than with either drug alone (Nelson et al., 1991). In addition, venlafaxine has a relatively safe side-effect profile allowing for rapid escalation of the dose if necessary (Derivan et al., 1995). If a rapid onset of action is a feature of venlafaxine then it may offer advantages over currently available antidepressant compounds by reducing patient suffering and thereby increasing compliance. In addition, the use of a single agent versus two or more agents to obtain equivalent antidepressant efficacy is desirable since it reduces the risk of drug-drug interactions (Andrews et al., 1996).

Mirtazapine

Mirtazapine enhances NA release *in vivo* by blockade of α_2 autoreceptors and serotonin release by blockade of α_2 heteroceptors and by increasing 5-HT cell firing (De Boer et al.,1994; De Boer et al.,1995; De Boer et al., 1996; Haddjeri et al., 1996). It has been suggested that its pharmacological profile is characterised by α_2-adrenergic, 5-HT$_2$, 5-HT$_3$ and histamine (H1) antagonist properties. It is devoid of anticholinergic activity and has no effect on monoamine re-uptake (Nickolson et al.,1992; De Boer et al., 1988). In some placebo-controlled studies, mirtazapine has been reported to exhibit antidepressant effects as early as after 1 week of treatment, with significant reductions on HAM-D scores and MADRS scores when compared to placebo (Smith et al., 1990; Bremner, 1995; Claghorn and Lesem, 1995), an improvement which is sustained throughout the treatment period (Burrows and Kremer, 1997).

BIMT 17

BIMT 17 is an antidepressant with 5-HT$_{1A}$ agonist and 5-HT$_{2A}$ receptor antagonist properties. It has been reported to be effective in a variety of animal models including the forced swim test in mice (Cesana et al., 1995), the olfactory bulbectomised rat (Borsini et al., 1997) and the Learned Helplessness paradigm (Borsini et al., 1997). In addition, BIMT 17 exhibited antidepressant like effects in the chronic mild stress model in mice after a single dose (D'Aquila et al., 1997), a model which is usually only sensitive to chronic antidepressant treatment. The preclinical profile of BIMT 17 supports the view that it could have a faster onset of action than currently available antidepressants, however clinical trials are necessary before any conclusions regarding speed on onset of BIMT 17 can be made.

Other agents

An early onset of antidepressant action has been reported for other compounds including roxindole (Benkert et al., 1992), reboxetine (Montgomery, 1997), amoxapine (Hekimian et al., 1978; Takahashi et al., 1979; Donlon et al., 1981; McNair et al., 1984) and adinazolam (Othmer and Othmer, 1988), although placebo controlled trials are limited and further investigation of these claims are necessary.

Conclusions

Although it is generally accepted that the onset of action of antidepressants is associated with a 2 to 3 week delay, due to receptor adaptation, this view does not fit with the original studies with imipramine (Kuhn, 1957). Some researchers have suggested that the delayed response is "something of a myth" (Angst, 1978; Willner, 1989). Improvement of depressive symptoms may emerge as early as after 5 days for drug trial responders, irrespective of whether they were receiving antidepressant drugs or placebo (Stassen et al., 1993). Others have noted that longer term improvement, at 6 and 20 weeks in depressed patients, was predicted from the improvement shown at 6 days (Parker et al., 1985). More recent meta-analysis of antidepressant drug trials have shown an onset of action within 1 week for fluoxetine (Tollefson and Holman, 1994) and mirtazapine (Kasper, 1995). The clinical data, coupled with the evidence from preclinical studies of rapid receptor adaptation, suggests that a fast onset of antidepressant action is attainable. Several issues however remain unresolved. From a clinical perspective, rapid onset with an individual agent is not achieved in every patient treated. The identification of responsive patients remains an elusive goal and is coupled to an understanding of the biological and psychological processes underlying depressive illness. To some extent it may also be related to features of the illness itself, melancholic patients being regarded as more responsive to antidepressants than non-melancholic patients (Rush and Weissenburger, 1994). At a fundamental level the causal relationship between biochemical changes and depression or between depression and biochemical changes remains a conundrum, with strong proponents on both sides of the argument. Leaving aside this dilemma, the nature of the biochemical paradigm itself, rooted in arguments about serotonergic and catecholaminergic receptor adaptation, is in danger of becoming a self-fulfilling prophecy. In this context the demonstration of time-dependent sensitisation of biological systems, including long-term receptor changes following acute doses of antidepressants, may lead to a new paradigm for the understanding of drug action (Antelman et al., 1997). The remarks of Stassen et al., (1993) "that antidepressants may not change the pattern (Gestalt) of the natural course of recovery from depression, but rather, they seem only to trigger recovery in a group of patients" seem prescient.

References

Andrews JM, Ninan PT, Nemeroff CB (1996) Venlafaxine: a novel antidepressant that has a dual mechanism of action. *Depression* 4 (2): 48–56

Angst J (1978) Time lag of antidepressant effect. In: *Depressive Disorders*. S Garattini (ed). Symposia Medica Hoechst 13, 468, Schattauer, Stuttgart

Antelman SM, Soares JG, Gershon S (1997) Time – dependent sensitisation – possible implications for clinical psychopharmacology. *Behavioural Pharmacology* 8: 505–514

Benkert O, Brunder G, Wetzel H (1992) Dopamine autoreceptor agonists in the treatment of schizophrenia and major depression. *Pharmacopsychiatry* 25: 254–260

Benkert O, Brunder G, Wetzel H, Hackett D (1996) A randomised double blind comparison of a rapidly escalating dose of venlafaxine and imipramine in inpatients with major depression and melancholia. *Journal of Psychiatric Research* 30 (6): 441–451

Blier P, Bergeron R (1995) Effectiveness of pindolol with selected antidepressant drugs in the treatment of major depression. *Journal of Clinical Psychopharmacology* 15: 217–222

Blier P, Bergeron R (1997) Early onset of therapeutic action in depression and greater efficacy of antidepressant treatments – are they related? *International Clinical Psychopharmacology* 12 (Suppl. 3): S21–S28

Blier P, Chaput Y, De Montigny C (1988) Long term 5-HT reuptake blockade, but not monoamine oxidase inhibition, decreases the function of terminal 5-HT autoreceptors: an electrophysiological study in the rat brain. *Naunyn Schmiedeberg's Archives of Pharmacology* 337: 246–254

Blier P, De Montigny C (1994) Current advances and trends in the treatment of depression. *Trends in Pharmacological Sciences* 15: 220–226

Bocchetta A, Bernardi F, Burrai C, Pedditzi M, Delzompo M (1993) A double blind study of L-sulpiride versus amitriptyline in lithium-maintained bipolar depressives. *Acta Psychiatrica Scandinavica* 88 (6): 434–439

Bolanos-Jimenez F, Manhaes de Castro R, Fillion G (1994) Effect of chronic antidepressant treatment on 5-HT$_{1B}$ presynaptic heteroreceptors inhibiting acetylcholine release. *Neuropharmacology* 33 (1): 77–81

Bordet R, Thomas P, Dupuis B (1998) Effect of pindolol on onset of action of paroxetine in the treatment of major depression : intermediate analysis of a double blind placebo controlled trial. *American Journal of Psychiatry* 155 (10): 1346–1351

Bornstein S (1979) Cross over trial comparing the antidepressant effects of amineptine and maprotiline. *Current Medical Research Opinions* 6 (2): 107–110

Borsini F, Cesana R, Kelly J, Leonard BE, McNamara M, Richards J, Seiden L (1997) BIMT 17: a putative antidepressant with a fast onset of action? *Psychopharmacology* 134: 378–386

Bremner JD (1995) A double blind comparison of ORG 3770, amitriptyline and placebo in major depression. *Journal of Clinical Psychiatry* 53 (11): 519–525

Briley M, Moret C (1993) 5-HT and antidepressants: *in vitro* and *in vivo* release studies. *Trends in Pharmacological Sciences* 14: 396–397

Broekkamp CLE, Leysen D, Peeters BW, Pinder RM (1995) Prospects for improved antidepressants. *Journal of Medicinal Chemistry* 38 (23): 4615–4633

Burnett FE, Dinan TG (1998) The Clinical Efficacy of Venlafaxine In The Treatment of Depression. *Reviews in Contemporary Pharmacotherapy* 9 (5): 303–320

Burrows GD, Kremer CME (1997) Mirtazapine – clinical advantages in the treatment of depression. *Journal of Clinical Psychopharmacology* 17 (2, Suppl. 1): S34–S39

Cervo L, Grignaschi G, Nowakowska E, Samanin R (1989) 1-(3 Trifluoromethylphenyl)-piperazine (TFMPP) in the ventral tegmental area reduces the effect of desipramine in the forced swimming test in rats : possible role of serotonin receptors. *European Journal of Pharmacology* 171 (1): 119–125

Cesana R, Ciprandi C, Bordsini F (1995) The effect of BIMT 17, a new potential antidepressant in the forced swim test in mice. *Behavioural Pharmacology* 6 (7): 688–694

Charney DS, Price LH, Heninger GR (1986) Desipramine-yohimbine combination treatment of refractory depression. Implications for the β-adrenergic receptor hypothesis of antidepressant action. *Archives of General Psychiatry* 43: 1156–1161

Claghorn JL, Lesem MD (1995) A double blind placebo controlled study of ORG 3770 in depressed outpatients. *Journal of Affective Disorders* 34 (3): 165–171

Crews FT, Paul SM, Goodwin FK (1981) Acceleration of β-receptor desensitisation by combined administration of antidepressants and phenoxybenzamine. *Nature* 290 (5809): 787–789

Dalery J, Rochat C, Peyron E, Bernard G (1992) Comparative study of the efficacy and acceptability of amineptine and fluoxetine in patients with major depression. *Encephale* 18 (3): 257–262

Dalery J, Rochat C, Peyron E, Bernard G (1997) The efficacy and acceptability of amineptine versus fluoxetine in major depression. *International Clinical Psychopharmacology* 12 (Suppl. 3): S35–S38

D'Aquila P, Monleon S, Borsini F, Brain P, Wilner P (1997) Anti-anhedonic actions of the novel serotonergic agent flibanserin, a potential antidepressant. *European Journal of Pharmacology*: 340, (2–3): 121–132

De Boer T, Maura G, Raitieri M, De Vos CJ, Wieringa J, Pinder RM (1988) Neurochemical and autonomic profiles of the 6-aza-analogue of mianserin, ORG 3770 and its enantiomers. *Neuropharmacology* 27: 399

De Boer T, Nefkens F, Van Helvoirt A, van Delft AML (1996) Differences in modulation of noradrenergic and serotonergic transmission by the alpha-2 adrenoceptor antagonists, mirtazapine, mianserin and idazoxan. *Journal of Pharmacology and Experimental Therapeutics*. 277: 852

De Boer T, Nefkens F, Van Helvoirt A (1994) The α_2-antagonist ORG 3770 enhances serotonin transmission *in vivo*. *European Journal of Pharmacology* 253 (1–2): R5–R6

De Boer T, Ruigt GSF, Berendsen HHG (1995) The α_2-selective adrenoceptor antagonist ORG 3770 (Mirtazapine, Remeron) enhances noradrenergic and serotonergic neurotransmission. *Human Psychopharmacology* 10 (Suppl. 2): S107–S119

Deniker P, Besancon G, Colonna L et al (1982) Extensive multicentric study of 1354 cases of depressed subjects treated with amineptine. *Encephale* 8: 355–370

Dennis T, Beauchemin V, Lavoie N (1994) Antidepressant – induced modulation of $GABA_A$ receptors and β-adrenoceptors but not $GABA_B$ receptors in the frontal cortex of olfactory bulbectomised rats. *European Journal of Pharmacology* 262 (1–2): 143–148

Derivan AT (1995) Antidepressants: can we determine how quickly they work? Issues from the literature. *Psychopharmacology Bulletin* 31: 23–28

Derivan A, Entsuh R, Kikta D (1995) Venlafaxine: measuring the onset of antidepressant action. *Psychopharmacology Bulletin* 31 (2): 439–447

Dinan TG, Scott LV (1996) Does pindolol induce a rapid improvement in depressed patients resistant to serotonin reuptake inhibitors? *Journal of Serotonin Research* 3: 119–121

Donlon PT, Biertuemphel H III, Willenbring M (1981) Amoxapine and amitriptyline in the outpatient treatment of endogenous depression. *Journal of Clinical Psychiatry* 38: 1048–1051

Edwards E, Harkins K, Wright G, Henn FA (1991) 5-HT_{1B} receptors in an animal model of depression. *Neuropharmacology* 30 (1): 101–105

Engel G, Gothert M, Hoyer D, Schlicker E, Hillenbrand K (1986) Identity of inhibitory presynaptic 5-HT autoreceptors in the rat brain cortex with 5-HT$_{1B}$ binding sites. *Naunyn-Schmiedeberg's Archives Pharmacology* 332 (1): 1–7

Entsuh R, Derivan A, Kikta D (1998) Early onset of antidepressant action of venlafaxine: pattern analysis in intent-to-treat patients. *Clinical Therapeutics* 20 (3): 517–526

Frances M, Khidichian F (1990) Chronic but not acute antidepressants interfere with serotonin (5-HT$_{1B}$) receptors. *European Journal of Pharmacology* 179: 173–176

Freeman H (1997) Early onset of action of amineptine. *International Clinical Psychopharmacology* 12 (Suppl. 3): S29–S33

Garcha G, Smokcum RWJ, Stephenson JD, Weeramanthri TB (1985) Effects of some atypical antidepressants on β-adrenoceptor binding and adenylate cyclase activity in the rat forebrain. *European Journal Pharmacology* 108: 1–7

Gessa GL (1996) Dysthymia and depressive disorders – dopamine hypothesis. *European Psychiatry* 11 (Suppl. 3): S123–S127

Greenberg PE, Stiglin LE, Finkelstein SN, Berndt ER (1993) Depression – a neglected major illness. *Journal of Clinical Psychiatry* 54 (1): 419–424

Guelfi JD, White C, Hackett D, Magni G (1995) Effectiveness of venlafaxine in patients hospitalised for major depression and melancholia. *J Clin Psychiatry* 56(10): 450–458

Haddjeri N, Blier P, De Montigny C (1996) Effect of the alpha-2-adrenoceptor antagonist mirtazapine on the 5-hydroxytryptamine system in the rat brain. *Journal of Pharmacology and Experimental Therapeutics* 277 (2): 861–871

Hekimian LJ, Friedhoff AJ, Deever E (1978) A comparison of the onset of action and therapeutic efficacy of amoxapine and amitriptyline. *Journal of Clinical Psychiatry* 39 (7): 633–637

Hervas I, Artigas F (1998) Effect of fluoxetine on extracellular 5-hydroxytrytamine in rat brain. Role of 5-HT autoreceptors. *European Journal of Pharmacology* 358: 9–18

Kasper S (1995) Clinical efficacy of mirtazapine : a review of meta-analyses of pooled data. *International Clinical Psychopharmacology* 10 (Suppl. 4): 25–35

Kemali D (1989) A Multicentre Italian Study Of Amineptine. *Clinical Neuropharmacology* 12 (Suppl. 2): S41–S50

Kendall DA, Duman R, Slopis J, Enna SJ (1982) Influence of adrenocorticotropin hormone and yohimbine on antidepressant-induced declines in rat brain neurotransmitter receptor binding function. *Journal of Pharmacology and Experimental Therapeutics* 222: 566

Kuhn R (1957) Über die Behandlung depressiver Zustände mit einem iminodibenzylderivat. *Schweizerische Medizinische Wochenschrift* 87: 1135–1140

Martin P, Puech AJ (1991) Is there a relationship between 5-HT$_{1B}$ receptors and the mechanisms of action of antidepressant drugs in the learned helplessness paradigm in rats. *European Journal of Pharmacology* 192 (1): 193–196

McGrath C (1996) The Olfactory bulbectomised rat model of depression: biochemical, neurochemical and pharmacological studies. Ph.D. Thesis. National University of Ireland

McNair D, Kahn RJ, Frankenthaler LM, Faldetta LL (1984) Amoxapine and amitriptyline I. Relative speed of antidepressant action. *Psychopharmacology* 83: 129–133

Mishra R, Sulser F (1978) Role of serotonin reuptake inhibition in the development of subsensitivity of the norepinephrine (NE) receptor coupled adenylate cyclase system. *Communications Psychopharmacology* 2 (4): 365–369

Montgomery SA (1992) "Evidence For Early Onset Of Action And Superior Efficacy. In: Controlled Trials With Venlafaxine. Presented at Venlafaxine: A New Dimension In Antidepressant Pharmacotherapy. Official satellite symposium to the XVIIIth CINP Congress, Nice, France, June 28

Montgomery SA (1997) Is there a role for a pure noradrenergic drug in the treatment of depression? *European Neuropsychopharmacology* 7 (Suppl. 1): S3–S9

Montgomery SA, Asberg M (1979) A new depression scale designed to be sensitive to change. *British Journal of Psychiatry* 134: 382–389

Moret C, Briley M (1990) Serotonin receptor subsensitivity and antidepressant activity. *European Journal of Pharmacology* 180: 351–356

Müller H, Möller HJ (1998) Methodological problems in the estimation of the onset of the antidepressant effect. *Journal of Affective Disorders* 48 (1): 15–23

Murray CLJ, Lopez JL (1996) The global burden of disease. Harvard University Press, Cambridge, UK

Nelson DR, Palmer KJ, Johnson AM (1990) Effect of prolonged 5-hydroxytryptamine uptake by paroxetine on cortical beta 1 and beta 2 adrenoceptors in rat brain. *Life Science* 47 (18): 1683–1691

Nickolson VJ, Wieringa JH, Van Delft AM (1982) Comparative pharmacology of mianserin, its main metabolites and 6-azamianserin. *Naunyn-Schmeidebergs Archives of Pharmacology* 319 (1): 48–55

Norman TR, Leonard BE (1994) Fast-acting antidepressants : can the need be met? *CNS Drugs* 2 (2): 120–131

Ollat H (1988) The mesolimbic dopamine system : a final common view on antidepressants. *Psychological Medicine* 20A: 5–9

Ordway GA, Gambarana C, Frazer A (1988) Quantitative autoradiography of central-adrenoceptor subtypes: comparison of the effects of chronic treatment with desipramine or centrally administered 1-isoproternol. *Journal of Pharmacology and Experimental Therapeutics* 247: 379

Othmer SC, Othmer E (1988) Adinazolam, a fast acting antidepressant, is tolerated during general anaesthesia (a letter). *Journal of Clinical Psychopharmacology* 8 (2): 151–152

Paes de Sousa M, Tropa J (1989) Evaluation of the efficacy of amineptine in a population of 1229 depressed patients: results of a multicenter study carried out by 135 general practitioners. *Clinical Neuropharmacology* 12 (Suppl. 2): S77–S86

Parker G, Tennant C, Blignault I (1985) Predicting improvement in patients with non-endogenous depression. *British Journal of Psychiatry* 146: 132–139

Pauwels PJ (1997) 5-$HT_{1B/D}$ Receptor Antagonists. *General Pharmacology* 29 (3): 293–303

Peroutka SJ (1993) 5-hydroxytrytamine receptors. *Journal of Neurochemistry* 60 (2): 408–416

Prien RF, Levine J (1984) Research and methodological issues for evaluating the therapeutic effectiveness of antidepressant drugs. *Psychopharmacology Bulletin* 20: 250–257

Roberts C, Belenguer A, Middlemiss DN, Routledge C (1998) Differential effects of 5-$HT_{1B/1D}$ receptor antagonists in dorsal and median raphe innervated brain regions. *European Journal of Pharmacology* 346: 175–180

Roberts C, Price GW, Jones BJ (1997) The role of 5-$HT_{1B/1D}$ receptors in the modulation of 5-hydroxytryptamine levels in the frontal cortex of the conscious guinea pig. *European Journal of Pharmacology* 326: 23–30

Rush AJ, Weissenburger JE (1994) Melancholic symptom features and DSM-IV. *American Journal of Psychiatry* 151: 489–498

Sachs GS, Pollack MH, Brotman AW, Farhadi AM, Gelenberg AJ (1986) Enhancement of ECT benefit by yohimbine. *Journal of Clinical Psychiatry* 47 (10): 508–510

Scott JA, Crews FT (1983) Rapid decrease in rat brain beta-adrenergic receptor binding during combined antidepressant alpha 2 antagonist treatment. *Journal of Pharmacology and Experimental Therapeutics* 224: 640–646

Sharp T, Umbers V, Gartside SE (1997) Effect of selective a 5-HT reuptake inhibitor in combination with 5-HT_{1A} and 5-HT_{1B} receptor antagonists on extracellular 5-HT in rat frontal cortex *in vivo*. *British Journal of Pharmacology* 121: 941–946

Smith WT, Glaudin V, Panagides J, Gilvary E (1990) Mirtazapine vs. amitriptyline vs. placebo in the treatment of major depressive disorder. *Psychopharmacology Bulletin* 26 (2): 191–196

Smith AL, Weissman MM (1992) In: Paykel E (ed). Handbook of Affective Disorders. Edinburgh: Churchill Livingstone 111–129

Stassen HH, Delini-Stula A, Angst J (1993) Time course of improvement under antidepressant treatment: a survival analytical approach. *European Neuropsychopharmacology* 3: 127–135

Sulser F, Vetulani J, Mobley PL (1978) Mode of action of antidepressant drugs. *Biochemical Pharmacology* 27 (3): 257–261

Takahashi R, Sakuma A, Hara T, Kazematsuri H, Mori A, Saito Y, Murasaki H, Ogucchi T, Sakurai Y, Yuzuriha T, Takemura M, Wurokawa H, Kurita H (1979) Comparison of efficacy of amoxapine and imipramine in a multi-clinic double blind study using the WHO schedule for a standard assessment of patients with depressive disorders. *Journal of International Medical Research* 7: 7–8

Tollefson GD, Holman SL (1994) How long to onset of antidepressant action : a meta analysis of patients treated with fluoxetine or placebo. *International Clinical Psychopharmacology* 9: 245–250

Van Amerongen P (1979) Double blind Clinical trial of the antidepressant action of amineptine. *Current Medical Research Opinions* 6 (2): 93–100

Vetulani J, Stawarz RJ, Dingell JV, Sulser F (1976) A possible common mechanism of action of antidepressant treatments. *Naunyn-Schmiedebergs Archives of Pharmacology* 293: 109

Wiech NL, Ursillo RC (1980) Acceleration of desipramine induced decrease of rat corticocerebral beta-adrenergic receptors by yohimbine. *Communications in Psychopharmacology* 4 (2): 95–100

Willner P (1989) Sensitisation of the action of antidepressant drugs. *Psychoactive Drugs* 407–453

Willner P (1997) The mesolimbic dopamine system as a target for rapid antidepressant action. *International Clinical Psychopharmacology* 12 (Suppl. 3): S7–S14

Wolfe BB, Harden TK, Sporn JR, Molinoff PB (1978) Presynaptic modulation of beta adrenergic receptors in cerebral cortex after treatment with antidepressants. *Journal of Pharmacology and Experimental Therapeutics* 207 (2): 446–457

Calcium signaling cascades, antidepressants and major depressive disorders

Ian A. Paul

Department of Psychiatry, University of Mississippi Medical Center, 2500 North State St. Jackson, MS 39216-4505, USA

Introduction

The discovery of effective antidepressants among the monoamine oxidase inhibitors and tricyclic compounds inhibiting monoamine uptake stimulated and drove research on the actions of antidepressants and the pathophysiology of major depression for nearly two decades (Stone, 1983). Newly discovered compounds with selectivity to antagonize the reuptake of serotonin have driven this research in recent years (Caldecott-Hazard et al., 1991; Charney, 1998; Maj et al., 1984). However, the response rate with any given antidepressant treatment has remained essentially unchanged since the early 1960's. Regardless of whether patients are prescribed monoamine oxidase inhibitors, tricyclic compounds or serotonin reuptake inhibitors, response rates hover between 60–65%. Moreover, in all cases, 3–6 weeks of drug administration is required to produce a therapeutic response, suggesting the agency of factors other than or in addition to the acute effects of these drugs (Charney, 1998; Oswald et al., 1972). Given the difficulties in maintaining patients in treatment after unsuccessful drug trials and the risk of self-injurious behavior and suicide until remission of depressive symptoms, the need for rapidly acting antidepressants with high response rates remains as pressing now as it was 40 years ago.

The failure to identify rapidly acting antidepressants which are effective in a large majority of depressed patients stems from the largely serendipitous nature of antidepressant treatment development. Rational drug development depends on understanding either the etiology of the targeted disease or the mechanism of therapeutic/palliative action of successful treatments. Drugs so designed are specifically targeted at points along an etiological or therapeutic cascade. In the case of major depressive disorders we clearly do not understand the etiology of the disease(s) and our understanding of the mechanism of therapeutic action is comparatively conjectural.

Antidepressants and monoaminergic neurotransmitters

Beginning with studies focused on monoaminergic actions of antidepressant treatments, considerable attention has been given to identifying the therapeutic mechanism of action of antidepressant treatments in order to design and test an hypothesis of the etiology of major depressive disorders. These have included studies of antidepressant effects on monoamine uptake and catabolism, receptor regulation and more recently synthetic enzyme regulation. A detailed account of these interactions is beyond the scope of this review and excellent reviews of the subject have recently been published (Duman et al., 1997; Rossby and Sulser, 1997). A very brief overview is provided below.

Initial studies of the prototypic tricyclic antidepressant (TCA), imipramine, by Hertting and coworkers revealed this compound to be a potent inhibitor of norepinephrine (NE) reuptake (Glowinski and Axelrod, 1964; Hertting et al., 1961). Further, imipramine and other TCAs have been shown to rapidly reduce the turnover of NE. Later studies of monoamine oxidase-inhibiting (MAOI) antidepressants indicated that these drugs, like TCAs, rapidly increase synaptic availability of NE and reduce its turnover. These observations formed the basis of the general hypothesis that major depressive disorder resulted from an insufficiency of NE in the synaptic cleft (Schildkraut et al., 1967, 1971). Similarly, Carlsson and coworkers demonstrated that antidepressants also potently inhibited the reuptake of serotonin (5HT). Later studies also demonstrated that these drugs reduced 5HT turnover (Carlsson et al., 1969a, b, 1968, 1969c). Indeed, later versions of the amine hypothesis incorporated these data to form a theory embracing both NE and 5HT uptake inhibition and subsequent reductions in neurotransmitter turnover as contributors to the mechanism of action of antidepressants (reviewed in Caldecott-Hazard et al., 1991; Maj et al., 1984; Oswald et al., 1972; Stone, 1983; Sugrue, 1985; Vetulani, 1991).

The effects of TCA and MAOI antidepressants on monoaminergic turnover are rapid, generally occurring with a few hours of drug administration. Moreover, steady-state plasma and brain levels of antidepressant drugs can be achieved in both humans and other animals over 72–96 h of standard dosing (Oswald et al., 1972; Stone, 1983). In contrast, antidepressants require chronic administration to produce significant clinical effects (Klein and Davis, 1969; Oswald et al., 1972; Vetulani, 1991). Further, drugs such as amphetamine and cocaine, while potent inhibitors of monoamine reuptake are not effective antidepressants. In addition, the so-called "atypical" antidepressants such as iprindole and alaproclate do not possess clear effects at monoaminergic terminals and yet are as effective as TCAs and MAOIs (reviewed in Vetulani, 1991). Thus, the acute effects of these compounds at monoaminergic synapses do not fully account for their mechanism of action (Oswald et al., 1972). These observations indicate that pharmacodynamic effects or neurobiological "adaptation" in response to repeated treatment rather than acute treatment effects are involved in the therapeutic response to antidepressants (Maj et al., 1984; Oswald et al., 1972; Stahl, 1998; Stone, 1983).

The search for adaptive neurobiological responses to antidepressant treatments has demonstrated changes in recognition, synthetic and catabolic/uptake sites in both noradrenergic and serotonergic neurons. Chronic but not acute treatment with antidepressant has been reported to modulate β-, α_1- and α_2-adrenoceptors (Banerjee et al., 1977; Okada et al., 1986; Vetulani et al., 1976; Vetulani and Sulser, 1975; Wolfe et al., 1978) as well as 5-HT$_2$ receptors (Bergstom and Kellar, 1979). Likewise, chronic antidepressant treatments modulate the activity of both tyrosine hydroxylase (Melia and Duman, 1991; Melia et al., 1992a, b; Nestler et al., 1990; Rosin et al., 1995; Yu and Boulton, 1990) and tryptophan hydroxylase (Briley et al., 1996; Greckshc et al., 1997; Huether et al., 1997; Knapp and Mandell, 1983; Sinei and Redfern, 1993). Moreover, recent studies indicate that antidepressants can modulate the activity and density of the transporters for both norepinephrine (Klimek et al., 1997; Ordway, 1997; Ordway et al., 1997; Zhu et al., 1998; Zhu and Ordway, 1997) and serotonin (Faludi et al., 1994; Mongeau et al., 1998). Taken together, these data indicate a significant role for adaptation of monoaminergic neurotransmitter systems, especially norepinephrine and serotonin in the actions of antidepressants.

Nonetheless, the critical sequence of adaptive neurobiological events required to effect antidepressive therapeutics remains unclear (Charney, 1998). One reason for this is that adaptation of noradrenergic or serotonergic neurons/receptors has not been linked to adaptive changes in behavior in response to antidepressant treatment (Charney, 1998; Stone, 1983). Consequently, it is not clear what dysfunction of monoaminergic neurons is required to produce depressive symptoms. However, monoaminergic neurons account for a relatively small proportion of neurons in the CNS. Far larger numbers of neurons use the inhibitory amino acid neurotransmitter, GABA or the excitatory amino acid neurotransmitter glutamate.

Antidepressants and excitatory amino acids

As early as 1982 there were reports suggesting that glutamate metabolism in depressed patients differed significantly from controls (Altamura et al., 1993; Kim et al., 1982b). More recent studies have indicated that these abnormalities can be normalized with chronic antidepressant treatments (Maes et al., 1998; Mauri et al., 1998). *In vitro* studies have shown that antidepressant drugs bind to NMDA receptors (Sills and Loo, 1988) and inhibit the binding of NMDA receptor ligands (Reynolds and Miller, 1988). Similarly, several laboratories have reported that antidepressants can modulate the release and/or reuptake of glutamate (Bouron and Chatton, 1999; Golembiowska and Zylewska, 1999; Kim et al., 1982a; McCaslin et al., 1992; Prikhozhan et al., 1990). Likewise, at least one group reported that conditions resulting in antidepressant-sensitive behavioral changes also disrupted long-term potentiation, a neurobiological process dependent upon release of glutamate (Shors et al., 1989).

Based on these findings, Trullas and Skolnick proposed that antagonists of the NMDA receptor would display antidepressant-like properties in acute, preclinical screening procedures such as the forced swim and tail suspension tests (Trullas and Skolnick, 1990; for an excellent review of this literature see Skolnick, 1999). Since their seminal observation, studies in several laboratories have demonstrated that functional antagonists of the NMDA receptor including ligands at the glutamate, glycine, polyamine and ionophore recognition sites are as efficacious as tricyclic antidepressants in preclinical antidepressant screening procedures in mice (Layer et al., 1995; Trullas, 1997; Trullas et al., 1991, 1989; Trullas and Skolnick, 1990) and rats (Cai and McCaslin, 1992; Maj et al., 1992a, c, d; Przegalinski et al., 1997; Wedzony et al., 1995). Likewise, chronic treatment with NMDA receptor antagonists results in chronic behavioral effects analogous to those produced by chronic antidepressant treatments in the chronic mild stress (Papp and Moryl, 1994), learned helplessness (Meloni et al., 1993; Mjellem et al., 1993), footshock-induced aggression (Maj et al., 1995; Ossowska et al., 1997; Przegalinski et al., 1997) and olfactory bulbectomy models (Kelly et al., 1997).

Congruent with the behavioral effects of chronic treatment with NMDA receptor antagonists are the observations that chronic, but not acute treatment with these drugs results in a reduction in the density of forebrain β-adrenoceptors (Maj et al., 1993; Paul et al., 1992) and 5-HT$_2$ receptors (Papp et al., 1994). Similarly, as observed with other antidepressants, administration of NMDA receptor antagonists to animals processed in the forced swim test results in a rapid down-regulation of forebrain β-adrenoceptors (Wedzony et al., 1995). Together with the data from animal analogs of major depression, these data provide strong support for the hypothesis that NMDA receptor antagonists have antidepressant properties.

Conversely, antidepressant drugs produce time- and dose-dependent changes in the radioligand binding properties of the NMDA receptor (Bernard et al., 1994; Kelly et al., 1997; Maj et al., 1991, 1992b; Massicotte et al., 1993; Mjellem et al., 1993; Nowak et al., 1993, 1998; Paul et al., 1993, 1994; Pilc and Legutko, 1995a, b; Skolnick et al., 1996; Watkins et al., 1998). These changes display nearly complete within-class generality and are not obtained with closely related non-antidepressant treatments (between-class specificity) (Paul et al., 1994). Thus, in both behavioral and biochemical screening procedures, antagonists at the NMDA receptor complex behave in a manner comparable to clinically active antidepressants.

Likewise, preclinical procedures such as the forced swim test and chronic mild stress paradigm which produce antidepressant-sensitive changes in behavior also produce alterations in the radioligand binding properties of

Table 1. Parameters of [^{3}H]CGP-39653 binding in frontal cortical homogenates from suicide victims and controls

Group	Specific binding (fmol/mg protein)	%High Affinity	IC$_{50\,(high)}$ (nM)	IC$_{50\,(low)}$ (mM)
Control	262 ± 12	45 ± 5	118 ± 53	0.97 ± 0.34
Suicide	219 ± 10*	27 ± 6*	307 ± 211	2.90 ± 0.94

Data were obtained as described in Methods and are presented as the mean ± SEM of 22 age- and *post-mortem* interval matched subjects per group. A one-site model was assumed (% High affinity = 0) unless the sum of squares of the model was significantly reduced by employing a two-site model (F-test, InPlot 4.03, San Diego, CA). * $-p < 0.05$ vs. control, two-tailed Student's *t*-test). Reprinted from (Nowak et al., 1995a).

the NMDA receptor opposite to those produced by antidepressants (Nowak et al., 1995b; Paul, 1997). In addition, antidepressant treatment in the chronic mild stress paradigm which normalizes sucrose consumption deficits also normalizes the radioligand binding properties of the NMDA receptor (Nowak et al., 1995b; Paul, 1997).

Moreover, NMDA receptor abnormalities are observed in human suicide victims (Nowak et al., 1995a). As shown in Table 1, in normal human frontal cortical (Brodmann Area 10) homogenates, glycine displaces the binding of the glutamate receptor antagonist [^{3}H]CGP-39653 in biphasic manner, similar to its effects in rodent tissue. In contrast, the high affinity component of glycine displacement of [^{3}H]CGP-39653 is significantly reduced in frontal cortical homogenates from the majority of suicide victims (Tab. 1). Moreover, specific binding of 10 nM [^{3}H]CGP-39653 was significantly lower (40 ± 14%) in cortices from suicide victims relative to that from age- and *post-mortem* interval-matched control subjects. No differences between control and suicide tissue were observed in the IC$_{50}$ for either component of glycine displacement of [^{3}H]CGP-39653 binding. In addition, no differences between groups were observed in either specific or glycine-displaceable [^{3}H]5,7-dichlorokynurenic acid (DCKA) binding or in specific, glycine- or glutamate-stimulated [^{3}H]dizocilpine binding.

These data lend significant support to the hypothesis that dysfunction of the NMDA receptor complex is involved in the psychopathology underlying human suicide. However, it must be recognized that major depressives comprise only about 50% of suicide victims. Further studies using tissue from suicide victims accompanied by psychiatric autopsy to determine the nature of premortem psychopathology will be needed to unequivocally link NMDA receptor dysfunction to major depressive disorder. In addition, it remains to be seen whether this dysfunction extends to either the strychnine-insensitive glycine binding site or the NMDA receptor-coupled ionophore. Nonetheless, these data represent a direct demonstration that changes in the NMDA receptor complex in frontal cortex attend human behaviors linked to major depression.

Most recently, Krystal and coworkers have demonstrated that intravenous administration of low doses of the NMDA receptor antagonist, ketamine, results in robust and rapid relief of depressive symptoms (Berman et al. 2000). These effects appear some 4 h after the disappearance of the transient dissociative effects of ketamine and continue to improve over three days post-infusion. Berman and coworkers noted an average reduction in Hamilton Depression Rating Scale scores of ~15 points and reported that these effects persisted for up to 2 weeks post-infusion in the absence of any other antidepressant treatment. While this small study must be replicated, the effects of ketamine in depressed humans are precisely those predicted by studies in animal screening paradigms and analogs of depression.

Antidepressants and voltage-dependent calcium channels

An extensive clinical literature suggests a role for calcium homeostasis in the pathophysiology of human depression (reviewed in Ortolano et al., 1983 and Pucilowski, 1997). For example, hypoparathyriodism has long been associated with affective disturbance (Clark et al., 1962; Cogan et al., 1978), and the etiology of these disturbances has been hypothesized to be related to errors in calcium metabolism. In addition, changes in serum or CSF calcium levels have been reported in patients with depression not directly associated with endocrine dysfunction (Carman et al., 1977; Carman and Wyatt, 1977, 1979; Cogan et al., 1978; Dubovsky, 1993; Dubovsky et al., 1989; Faragalla and Flach, 1970; Flach, 1964, 1966; Flach et al., 1960; Gour and Chaudry, 1957; Jimerson et al., 1979; Levine et al., 1999; Linder et al., 1989). Conversely, reduced plasma, serum or CSF calcium levels accompany improvement of depressive symptomatology after either electroconvulsive shock or antidepressant drug treatment (Carman et al., 1977; Eiduson et al., 1960; Faragalla and Flach, 1970; Flach, 1964; Flach et al., 1960; Linder et al., 1989; Mellerup et al., 1979). Finally, calcium channel antagonists have been reported to be antidepressant in an anecdotal report (Dubovsky et al., 1985).

The hypothesis that alterations in intracellular calcium concentrations can affect performance in the forced swim test is consistent with several other lines of evidence. Thus, elevated calcium intake has been reported to enhance learned helplessness in rats (Trulson et al., 1986). Likewise, the dihydropyridine calcium channel agonist, BAY K 8644 enhances immobility in the forced swim test (Mogilnicka et al., 1988) and antagonizes the effects of IMI (Martin et al., 1989). In contrast, dihydropyridine calcium channel antagonists reduce the duration of immobility in the forced swim test in mice (Czyrak et al., 1989; Kostowski et al., 1990; Mogilnicka et al., 1987; Skolnick et al., 1992) and rats (Czyrak et al., 1989; Martin et al., 1989). Moreover, low doses of tricyclic antidepressants facilitate the actions of subeffective doses of dihydropyridine calcium channel antagonists (Martin et al., 1989; Mogilnicka et al., 1987).

These data clearly indicate that major depressive disorders are accompanied by abnormalities in calcium homeostasis. Moreover, there is considerable evidence for the use of L-type calcium channel antagonists as either antidepressants or antidepressant adjuncts. As with the linkage between excitatory amino acid neurotransmission, major depression and antidepressant activity, addition studies, particularly in clinical populations will be required to define this role. Nonetheless, the data to date provide compelling evidence for a role for calcium homeostasis in affective disorders and antidepressant treatments.

Excitatory amino acid and nitric oxide neurotransmission

The NMDA receptor is a ligand-gated, voltage-sensitive ionophore which gates calcium (Ca^{2+}) and, to a lesser extent, sodium (Na^+) and potassium (K^+) (Meguro et al., 1992). Stimulation of the receptor and opening of the ionophore results in Ca^{2+} entry into the receptive neuron. The Ca^{2+} binds to and stimulates a Ca^{2+}-calmodulin complex which in turn stimulates nitric oxide (NO) synthesis (Southam and Garthwaite, 1993) to convert L-arginine to L-citrulline and liberate NO. The released NO can then stimulate an NO-sensitive guanylyl cyclase to convert GTP to cGMP (Fig. 1). Alternately, NO can serve as a transsynaptic signal either at the presynaptic neuron or nearby cells. NO synthase, the enzyme responsible for the production of NO can be exogenously stimulated with compounds such as sodium nitroprusside, molsidomine, S-nitroso-N-acetylpenicillamine and peripherally administered L-arginine. Conversely, NO synthase activity can be inhibited with N^G-nitro-L-arginine (L-NNA), N^G-nitro-L-arginine methyl ester (L-NAME) and N^G-monomethyl-L-arginine (L-NMMA).

In addition to its effects on guanylyl cyclase, NO is a membrane-permeable molecule involved in signaling process and cellular communication in a variety of systems. Previous studies have suggested that NO may be able to stimulate the release of neurotransmitters (dopamine, norepinephrine) (Zhu and Luo, 1992) and this effect is dependent on NMDA receptor stimulation. In fact, both an increase (Strasser et al., 1994; Zhu and Luo, 1992) and decrease (Bowyer et al., 1995; Lin et al., 1995) in basal monoamine efflux after pretreatment with NO precursors or donors have been reported. Recent reports also indicate that NO plays the role of an inhibitory endogenous substance in discriminative effects of the psychostimulants in rats, because inhibition of NOS enhances the effects of amphetamine and cocaine while an increased NO level attenuates them (Filip and Przegalinski, 1998). Moreover microdialysis data have shown that NO has an inhibitory influence on the DA release in the stratum (Guevara-Guzman et al., 1994). Likewise Silva et al., (Silva et al., 1995)

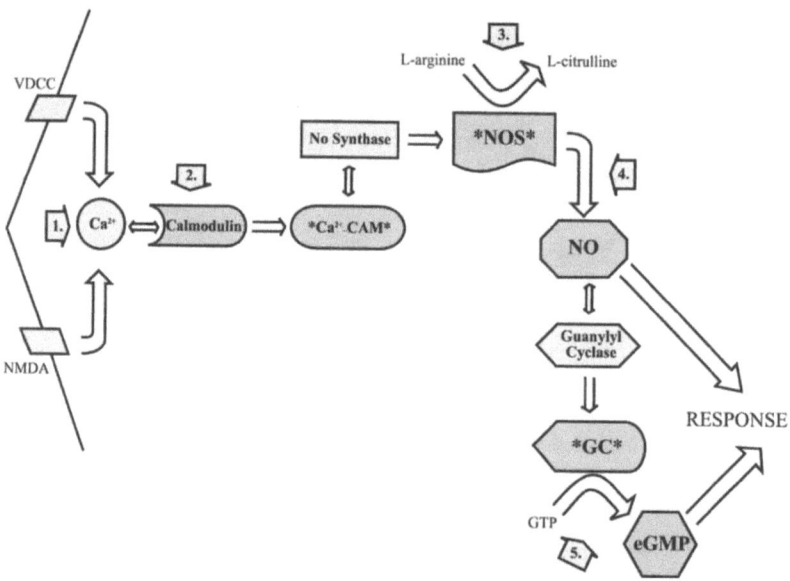

Figure 1. Ca^{2+}-calmodulin-nitric oxide synthase-guanylyl cyclase (CC-NOS-GC) Cascade. Abbreviations: VDCC = voltage-dependent calcium channel. NMDA = N-methyl-D-aspartate receptor. CAM = calmodulin. NO = nitric oxide. NOS = nitric oxide synthase. GC = guanylyl cyclase. GTP = guanosine triphosphate. cGMP = cyclic guanosine monophosphate. Abbreviations bracketed by asterisks (e.g. *Ca^{2+}-CAM*) denote activated form of enzymes. Numbered arrows (antagonists/agonists): 1. Calcium chelators/$CaCO_3$; 2. Calmodulin antagonists/$CaCO_3$; 3. Nitric oxide synthase antagonists/L-arginine; 4. NO "trapping" agents/molsidomine; 5. Guanylyl cyclase inhibitors/YC-1.

demonstrated an increased DA release *in vivo* following intrastriatal administration of 7-nitroindazole (7NI), which was antagonized by coperfusion with the NOS substrate, L-arginine. NMDA-mediated neurotransmitter release is linked to NO production and the NOS inhibitors L-NNA and 7-nitroindazole (7NI) blocked the NMDA-mediated release of neurotransmitters in dose-dependent fashion (Montague et al., 1994).

Antidepressants and nitric oxide

A potential role for NO in affective disorder has also recently been proposed (Harvey, 1996). NO is a regulator of both short- and long-term neuronal adaptive changes and, consequently, may play a role in neuronal adaptation to antidepressant drugs. Targets of NO include guanylate cyclase, G proteins, amino acid, amine and neuropeptide release and transport. NO mediated cGMP synthesis also mediates induction of immediate early gene expression which are implicated in long-term synaptic changes and more recently in the mechanism of action of antidepressant drugs (reviewed by (Harvey, 1996); see also (Duman et al., 1997)). Inhibitors of NO synthase are beginning to be developed for the various isoforms of NO synthase, which raises the possibility of developing well tolerated and selective NO synthase inhibitors that may be used for the treatment of CNS disorders where NO is implicated.

The first evidence of biochemical linkage between antidepressant medication and NOS activity was described by Finkel et al. (Finkel et al., 1996). They showed that the serotonin reuptake inhibitor, paroxetine, inhibits constitutive nitric oxide synthase (cNOS) activity in animal models and humans (i.e. at Site 3, Fig. 1) at concentrations comparable to those achieved in clinical therapy. NOS inhibition could theoretically cause hypertension, coronary artery spasm and platelet aggregation and thus be inappropriate for patients with ischemic heart disease and/or coronary insufficiency (Moncada and Higgs, 1993). However, paroxetine has an extraordinary safety profile and all examined patients with ischemic heart disease tolerated this NOS inhibitor without any serious adverse cardiac events.

We have recently demonstrated that antagonists of NO synthase are active in both acute and chronic preclinical antidepressant screening procedures (Harkin et al., 1999; Karolewicz et al., 1999). Thus, acute administration of NO synthase antagonists produces a robust reduction in immobility in the forced swim test comparable to that of the tricyclic, imipramine (Tab. 2). This effect is dose-dependent, stereoselective and

Table 2. Effect of nitric oxide synthase antagonists in the mouse forced swim test

Drug	Dose mg/kg	Duration of immobility, s.	% Control immobility
Saline	–	138 ± 12	100
Imipramine	15	74 ± 19	53*
	1	122 ± 19	88
	3	92 ± 11	66*
L-NNA	10	86 ± 10	63*
	30	126 ± 9	91
Saline	–	130 ± 13	100
	3	128 ± 14	99
D-NNA	10	142 ± 12	109
	30	138 ± 13	106
	3	146 ± 13	112
L-NMMA	10	100 ± 12	77
	30	76 ± 15	59*
	30	128 ± 13	99
	100	93 ± 11	72*
L-NAME	175	132 ± 14	102
	300	179 ± 10	138*

Data represent the mean ± S.E.M of 10 mice/group. *$p < 0.01$ Fisher's LSD following single factor ANOVA, (from Harkin et al., 1999).

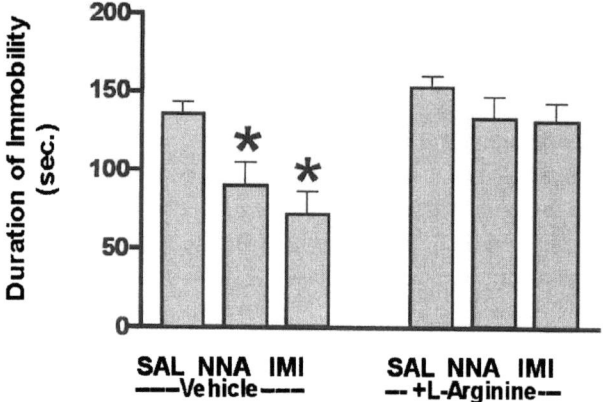

Figure 2. L-arginine (LA-500 mg/kg) reversal of the effects of N^G-nitro-L-arginine (NNA – 10 mg/kg) and imipramine (IMI – 15 mg/kg) in the forced swim test. Data represent the mean ± S.E.M of 9–10 subjects per group. *$p < 0.05$ vs. Saline, Fisher's LSD, (from Harkin et al., 1999).

reversed by cotreatment with the NO synthase substrate, L-arginine (Tab. 2, Fig. 2).

Likewise, chronic, but not acute treatment with an NO synthase antagonist results in a reduction in the density of cortical β-adrenoceptors in mice (Tab. 3). This effect is also dose-dependent and results in a U-shaped response comparable to that observed in the forced swim test (Tab. 2). Notably, L-NNA is as efficacious as imipramine at down-regulating β-adrenoceptors suggesting the possibility of comparable clinical efficacy.

Together, these data represent compelling evidence that the calcium signaling cascade beginning at the NMDA receptor and proceeding at least through the activation of NO synthase plays a significant role in affective behavior and the response to antidepressant treatments. We can now hypothesize: 1) that interruption of the Ca^{2+}-calmodulin-NOS-guanylyl cyclase

Table 3. The effect of chronic imipramine and L-NARG treatment on β-adrenoceptors in mouse cortex

Drug (mg/kg, i.p.)	Bmax (fmol/mg protein)	Kd (nM)
Saline	140 ± 14	0.25 ± 0.02
Imipramine 15	98 ± 5*	0.26 ± 0.01
Imipramine 30	106 ± 9*	0.25 ± 0.01
L-NNA 0.1	101 ± 7*	0.25 ± 0.01
L-NNA 0.3	97 ± 5*	0.25 ± 0.01
L-NNA 1	111 ± 8	0.25 ± 0.01

Values represent the mean ± SEM of 9–16 subject/group. Bmax and Kd values were determined using LIGAND program Version 5.0.3. Data were analyzed using a one-way ANOVA followed by Student-Neuman-Keuls and LSD *post-hoc* determination. *$p < 0.05$ vs. saline control, (from Karolewicz et al., 1999).

subcellular signaling pathway at any point will produce antidepressant-like effects; 2) that the acute actions of antidepressants in preclinical screening procedures are a consequence of their ability to disrupt Ca^{2+}-calmodulin-NOS-guanylyl cyclase signaling; 3) that chronic, but not acute treatment with antidepressants results in adaptation of the Ca^{2+}-calmodulin-NOS-guanylyl cyclase signaling pathway; 4) that this adaptation is necessary for the achievement of the therapeutic actions of antidepressants and; 5) that major depression is accompanied by an alteration (hyperactivity?) of subcellular Ca^{2+} signaling.

Although many more studies will be needed, the Ca^{2+}-calmodulin-NOS-guanylyl cyclase subcellular signaling pathway represents a hitherto unexploited means of understanding affective disorders and developing novel antidepressant treatments. Studies addressing these hypotheses will provide critical information regarding the role of the Ca^{2+}-calmodulin-NOS-guanylyl cyclase subcellular signaling pathway in major depression and the possibility of designing novel and effective antidepressant treatments based on interactions along this pathway.

Acknowledgement
The author gratefully acknowledges Drs. B. Karolewicz and A. Harkin for allowing the reproduction of figures from their manuscripts.

References

Altamura CA, Mauri MC, Ferrara A, Moro AR, D'Andrea G, Zamberlan F (1993) Plasma and platelet excitatory amino acids in psychiatric disorders. *Neuroscience* 55: 511–519

Banerjee SP, Kung LS, Riggi SJ, Chanda SK (1977) Development of beta adrenergic receptor subsensitivity by antidepressants. *Nature* 268: 455–456

Bergstom DA, Kellar KJ (1979) Effect of electoconvulsive shock on monoaminergic receptor binding in rat brain. *Nature* 278: 464–466

Berman RM, Cappiello A, Anand A, Oren DA, Heninger GR, Charney DS and Krystal JH (2000) Antidepressant effects of ketamine in depressed patients. *Biol Psychiatry* 47: 351–354

Bernard J, Ohayon M, Massicotte G (1994) Modulation of the AMPA receptor by phospholipase A2: effect of the antidepressant trimipramine. *Psychiatry Res* 51 (2): 107–114

Bouron A, Chatton JY (1999) Acute application of the tricyclic antidepressant desipramine presynaptically stimulates the exocytosis of glutamate in the hippocampus. *Neuroscience* 90 (3): 729–736

Bowyer JF, Clausing P, Gough B, Slikker W Jr, Holson RR (1995) Nitric oxide regulation of methamphetamine-induced dopamine release in caudate/putamen. *Brain Res* 699 (1): 62–70

Briley M, Prost JF, Moret C (1996) Preclinical pharmacology of milnacipran. *Int Clin Psychopharmacol* 11 Suppl 4: 9–14

Cai Z, McCaslin PP (1992) Amitriptyline, desipramine, cyproheptadine and carbamazepine, in concentrations used therapeutically, reduce kainate- and N-methyl-D-aspartate-induced intracellular Ca^{2+} levels in neuronal culture. *Eur J Pharmacol* 219 (1): 53–57

Caldecott-Hazard S, Morgan DG, DeLeon-Jones F, Overstreet DH, Janowsky D (1991) Clinical and biochemical aspects of depressive disorders – II. Transmitter/receptor theories. *Synapse* 9: 251–301

Carlsson A, Corrodi H, Fuxe K, Hokfelt T (1969a) Effect of antidepressant drugs on the depletion of intraneuronal brain 5-hydroxytryptamine stores caused by 4-methyl-alpha-ethyl-meta-tyramine. *Eur J Pharmacol* 5 (4): 357–366

Carlsson A, Corrodi H, Fuxe K, Hokfelt T (1969b) Effects of some antidepressant drugs on the depletion of intraneuronal brain catecholamine stores caused by 4,alpha-dimethyl-meta-tyramine. *Eur J Pharmacol* 5 (4): 367–373

Carlsson A, Fuxe K, Ungerstedt U (1968) The effect of imipramine on central 5-hydroxytryptamine neurons. *J Pharm Pharmacol* 20 (2): 150–151

Carlsson A, Jonason J, Lindqvist M, Fuxe K (1969c) Demonstration of extraneuronal 5-hydroxytryptamine accumulation in brain following membrane-pump blockade by chlorimipramine. *Brain Res* 12 (2): 456–460

Carman JS, Post RM, Goodwin FK, Bunney WE (1977) Calcium and electroconvulsive therapy in severe depressive illness. *Biol Psychiatry* 12: 5–17

Carman JS, Wyatt RJ (1977) Alterations in cerebrospinal fluid and serum total calcium with changes in psychiatric state. In: E Usdin et al (eds). Neuroregulators and psychiatric disorders. New York, Oxford Univ Press

Carman JS, Wyatt RJ (1979) Calcium: Bivalent cation in the bivalant psychoses. *Biol Psychiatry* 14: 295–336

Charney DS (1998) Monoamine dysfunction and the pathophysiology and treatment of depression. *J Clin Psychiatry* 59 Suppl 14: 11–14

Clark JA, Davidson LJ, Ferguson HC (1962) Psychosis in hypoparathyroidism. *J Mental Sci* 108: 811–815

Cogan MG, Covey CM, Arieff AI, Wisniewski A, Clark OH, Lazarowitz V, Leach W (1978) Central nervous system manifestations of hyperparathyroidism. *Am J Med* 65 (6): 963–970

Czyrak A, Mogilnicka E, Maj J (1989) Dihydropyridine calcium channel antagonists as antidepressant drugs in mice and rats. *Neuropharmacology* 3: 229–233

Dubovsky SL (1993) Calcium Antagonists in Manic-Depressive Illness. *Neuropsychobiology* 27: 184–192

Dubovsky SL, Christiano J, Daniell LC, Franks RD, Murphy J, Adler L, Baker N, Harris RA (1989) Increased platelet intracellular calcium concentration in patients with bipolar affective disorders. *Arch Gen Psychiatry* 46 (7): 632–638

Dubovsky SL, Franks RD, Schrier D (1985) Phenelzine-induced hypomania: Effect of verapamil. *Biol Psychiatry* 20: 1009–1014

Duman RS, Heninger GR, Nestler EJ (1997) A molecular and cellular theory of depression. *Arch Gen Psychiatry* 54 (7): 597–606

Eiduson S, Brill NQ, Crumpton E (1960) The effect of electro-convulsive therapy on spinal fluid constituents. *J Mental Sci* 106: 692–698

Faludi G, Tekes K, Tothfalusi L (1994) Comparative study of platelet 3H-paroxetine and 3H-imipramine binding in panic disorder patients and healthy controls. *J Psychiatry Neurosci* 19 (2): 109–113

Faragalla FF, Flach FF (1970) Studies of mineral metabolism in mental depression. I. The effects of imipramine and electric convulsive therapy on calcium balance and kinetics. *J Nerv Ment Dis* 151 (2): 120–129

Filip M, Przegalinski E (1998) The role of the nitric oxide (NO) pathway in the discriminative stimuli of amphetamine and cocaine. *Pharmacol Biochem Behav* 59 (3): 703–708

Finkel MS, Laghrissi-Thode F, Pollock BG, Rong J (1996) Paroxetine is a novel nitric oxide synthase inhibitor. *Psychopharmacol Bull* 32 (4): 653–658

Flach FF (1964) Calcium metabolism in states of depression. *Br J Psychiatr* 110: 588–593

Flach FF (1966) The impact of pharmacotherapy on psychiatric practice and research. *Ment Hyg* 50 (4): 570–573

Flach FF, Liang E, Stokes PE (1960) The effects of electic convulsive treatments on nitrogen, calcium, and phosphorus. *J Mental Sci* 106: 638–647

Glowinski J, Axelrod J (1964) Inhibition of tritiated noradrenaline in intact rat brain by imipramine and structurally related compounds. *Nature* 204: 1318–1319

Golembiowska K, Zylewska A (1999) Effect of antidepressant drugs on veratridine-evoked glutamate and aspartate release in rat prefrontal cortex. *Pol J Pharmacol* 51 (1): 63–70

Gour KN, Chaudry HM (1957) Study of calcium metabolism in electric convulsive therapy (ECT) in certain. *J Mental Sci* 103: 275–285

Greckschh G, Zhou D, Franke C, Schroder U, Sabel B, Becker A, Huether G (1997) Influence of olfactory bulbectomy and subsequent imipramine treatment on 5-hydroxytryptaminergic presynapses in the rat frontal cortex: behavioural correlates. *Br J Pharmacol* 122 (8): 1725–1731

Guevara-Guzman R, Emson PC, Kendrick KM (1994) Modulation of *in vivo* striatal transmitter release by nitric oxide and cyclic GMP. *J Neurochem* 62 (2): 807–810

Harkin AJ, Bruce KH, Craft B, Paul IA (1999) Nitric oxide synthase inhibitors have antidepressant-like properties in mice. 1. Acute treatments are active in the forced swim test. *Eur J Pharmacol* 372: 207–213

Harvey BE (1996) Affective disorders and nitric oxide: A role in pathways to relapse and refractoriness? *Human Psychopharmacology* 11: 309–319

Hertting G, Axelrod J, Whitby LG (1961) Effect of drugs on the uptake and metabolism of H^3-norepinephrine. *J Pharmacol Exp Ther* 134: 146–153

Huether G, Zhou D, Ruther E (1997) Long-term modulation of presynaptic 5-HT-output: experimentally induced changes in cortical 5-HT-transporter density, tryptophan hydroxylase content and 5-HT innervation density. *J Neural Transm* 104 (10): 993–1004

Jimerson DC, Post RM, Carman JS, van Kammen DP, Wood JH, Goodwin FK, Bunney WE (1979) CSF Calcium: Clinical corellates in affective illness and schizophrenia. *Biol Psychiatry* 14: 37–51

Karolewicz B, Bruce KH, Lee B, Paul IA (1999) Nitric oxide synthase inhibitors have antidepressant-like properties in mice. 2. Chronic treatment results in downregulation of cortical β-adrenoceptors. *Eur J Pharmacol* 372: 215–220

Kelly JP, Wrynn AS, Leonard BE (1997) The olfactory bulbectomized rat as a model of depression: an update. *Pharmacol Ther* 74 (3): 299–316

Kim JS, Schmid-Burgk W, Claus D, Kornhuber HH (1982a) Effects of amitriptyline on serum glutamate and free tryptophan in rats. *Arch Psychiatr Nervenkr* 232 (5): 391–394

Kim JS, Schmid-Burgk W, Claus D, Kornhuber HH (1982b) Increased serum glutamate in depressed patients. *Arch Psychiatr Nervenkr* 232 (4): 299–304

Klein DF, Davis JM (1969) Diagnosis and treatment of psychiatric disorders, Baltimore,MD, The Williams and Wilkins Co.

Klimek V, Stockmeier C, Overholser J, Meltzer HY, Kalka S, Dilley G, Ordway GA (1997) Reduced levels of norepinephrine transporters in the locus coeruleus in major depression. *J Neurosci* 17 (21): 8451–8458

Knapp S, Mandell AJ (1983) Lithium and chlorimipramine differentially alter the stability properties of tryptophan hydroxylase as seen in allosteric and scattering kinetics. *Psychiatry Res* 8 (4): 311–323

Kostowski W, Dyr W, Pucilowski O (1990) Activity of diltiazem and nifedipine in some animal models of depression. *Pol J Pharmacol Pharm* 42 (2): 121–127

Layer RT, Popik P, Olds T, Skolnick P (1995) Antidepressant-like actions of the polyamine site NMDA antagonist, eliprodil (SL-82.0715). *Pharmacol Biochem Behav* 52 (3): 621–627

Levine J, Stein D, Rapoport A, Kurtzman L (1999) High serum and cerebrospinal fluid Ca/Mg ratio in recently hospitalized acutely depressed patients. *Neuropsychobiology* 39 (2): 63–70

Lin AM, Kao LS, Chai CY (1995) Involvement of nitric oxide in dopaminergic transmission in rat striatum: an *in vivo* electrochemical study. *J Neurochem* 65 (5): 2043–2049

Linder J, Brismar K, Beck-Friis J, Saaf J, Wetterberg L (1989) Calcium and magnesium concentrations in affective disorder: difference between plasma and serum in relation to symptoms. *Acta Psychiatr Scand* 80 (6): 527–537

Maes M, Verkerk R, Vandoolaeghe E, Lin A, Scharpe S (1998) Serum levels of excitatory amino acids, serine, glycine, histidine, threonine, taurine, alanine and arginine in treatment-resistant depression: modulation by treatment with antidepressants and prediction of clinical responsivity. *Acta Psychiatr Scand* 97 (4): 302–308

Maj J, Klimek V, Golembiowska K, Rogoz Z, Skuza G (1993) Central effects of repeated treatment with CGP 37849, a competitive NMDA receptor antagonist with potential antidepressant activity. *Pol J Pharmacol* 45 (5–6): 455–466

Maj J, Przegalinski E, Mogilnicka E (1984) Hypotheses concerning the mechanism of action of antidepressant drugs. *Rev Physiol Biochem Pharmacol* 100: 1–74

Maj J, Rogoz Z, Skuza G (1991) Antidepressant drugs increase the locomotor hyperactivity induced by MK-801 in rats. *J Neural Transm Gen Sect* 85 (3): 169–179

Maj J, Rogoz Z, Skuza G (1992a) The effects of combined treatment with MK-801 and antidepressant drugs in the forced swimming test in rats. *Pol J Pharmacol Pharm* 44 (3): 217–226

Maj J, Rogoz Z, Skuza G, Kolodziejczyk K (1995) Some central effects of GYKI 52466, a non-competitive AMPA receptor antagonist. *Pol J Pharmacol* 47 (6): 501–507

Maj J, Rogoz Z, Skuza G, Sowinska H (1992b) The effect of antidepressant drugs on the locomotor hyperactivity induced by MK-801, a non-competitive NMDA receptor antagonist. *Neuropharmacology* 31 (7): 685–691

Maj J, Rogoz Z, Skuza G, Sowinska H (1992c) The effect of CGP 37849 and CGP 39551, competitive NMDA receptor antagonists, in the forced swimming test. *Pol J Pharmacol Pharm* 44 (4): 337–346

Maj J, Rogoz Z, Skuza G, Sowinska H (1992d) Effects of MK-801 and antidepressant drugs in the forced swimming test in rats. *Eur Neuropsychopharmacol* 2 (1): 37–41

Martin P, Laurent S, Massol J, Childs M, Puech AJ (1989) Effects of dihydropyridine drugs on reversal by imipramine of helpless behavior in rats. *Eur J Pharmacol* 162: 185–188

Massicotte G, Bernard J, Ohayon M (1993) Chronic effects of trimipramine, an antidepressant, on hippocampal synaptic plasticity. *Behav Neural Biol* 59 (2): 100–106

Mauri MC, Ferrara A, Boscati L, Bravin S, Zamberlan F, Alecci M, Invernizzi G (1998) Plasma and platelet amino acid concentrations in patients affected by major depression and under fluvoxamine treatment. *Neuropsychobiology* 37 (3): 124–129

McCaslin PP, Yu XZ, Ho IK, Smith TG (1992) Amitriptyline prevents N-methyl-D-aspartate (NMDA)-induced toxicity, does not prevent NMDA-induced elevations of extracellular glutamate, but augments kainate-induced elevations of glutamate. *J Neurochem* 59 (2): 401–405

Meguro H, Mori H, Araki K, Kushiya E, Kutsuwada T, Yamazaki M, Kumanishi T, Arakawa M, Sakimura K, Mishina M (1992) Functional characterization of a heteromeric NMDA receptor channel from cloned cDNAs. *Nature* 357: 70–74

Melia KR, Duman RS (1991) Involvement of corticotropin-releasing factor in chronic stress regulation of the brain noradrenergic system. *Proc Natl Acad Sci* 88: 8382–8386

Melia KR, Nestler EJ, Duman RS (1992a) Chronic imipramine treatment normalizes levels of tyrosine hydroxylase in the locus coeruleus of chronically stressed rats. *Psychopharmacology* (Berl) 108 (1–2): 23–26

Melia KR, Rasmussen K, Terwilliger RZ, Haycock JW, Nestler EJ, Duman RS (1992b) Coordinate regulation of the cyclic AMP system with firing rate and expression of tyrosine hydroxylase in the rat locus coeruleus: effects of chronic stress and drug treatments. *J Neurochem* 58 (2): 494–502

Mellerup ET, Bech P, Sorensen T, Frederiksen AF, Rafaelsen OJ (1979) Calcium and electroconvulsive therapy of depressed patients. *Biol Psychiatry* 14 (4): 711–714

Meloni D, Gambarana C, De Montis MG, Dal Pra P, Taddei I, Tagliamonte A (1993) Dizocilpine antagonizes the effect of chronic imipramine on learned helplessness in rats. *Pharmacol Biochem Behav* 46 (2): 423–426

Mjellem N, Lund A, Hole K (1993) Reduction of NMDA-induced behaviour after acute and chronic administration of desipramine in mice. *Neuropharmacology* 32 (6): 591–595

Mogilnicka E, Czyrak A, Maj J (1987) Dihydropyridine calcium channel antagonists reduce immobility in the mouse behavioral despair teat; antidepressants facilitate nifedipine action. *Eur J Pharmacol* 138: 413–416

Mogilnicka E, Czyrak A, Maj J (1988) BAY K 8644 enhances immobility in the mouse behavioral despair test, an effect blocked by nifedipine. *Eur J Pharmacol* 151: 307–311

Moncada S, Higgs A (1993) The L-arginine-nitric oxide pathway. *N Engl J Med* 329: 2002–2012

Mongeau R, Weiss M, de Montigny C, Blier P (1998) Effect of acute, short- and long-term milnacipran administration on rat locus coeruleus noradrenergic and dorsal raphe serotonergic neurons. *Neuropharmacology* 37 (7): 905–918

Montague PR, Gancayco CD, Winn MJ, Marchase RB, Friedlander MJ (1994) Role of NO production in NMDA receptor-mediated neurotransmitter release in cerebral cortex. *Science* 263 (5149): 973–977

Nestler EJ, McMahon A, Sabban EL, Tallman JF, Duman RS (1990) Chronic antidepressant administration decreases the expression of tyrosine hydroxylase in the rat locus coeruleus. *Proc Natl Acad Sci* 87: 7522–7526

Nowak G, Legutko B, Skolnick P, Popik P (1998) Adaptation of cortical NMDA receptors by chronic treatment with specific serotonin reuptake inhibitors. *Eur J Pharmacol* 342 (2–3): 367–370

Nowak G, Ordway GA, Paul IA (1995a) Alterations in the N-methyl-D-aspartate (NMDA) receptor complex in the frontal cortex of suicide victims. *Brain Res* 675: 157–164

Nowak G, Redmond A, McNamara M, Paul IA (1995b) Swim stress increases the potency of glycine at the N-methyl-D-aspartate receptor complex. *J Neurochem* 64 (2): 925–927

Nowak G, Trullas R, Layer RT, Skolnick P, Paul IA (1993) Adaptive changes in the N-methyl-D-aspartate receptor complex after chronic treatment with imipramine and 1-aminocyclopropanecarboxylic acid. *J Pharmacol Exp Ther* 265 (3): 1380–1386

Okada F, Tokumitsu Y, Ui M (1986) Desensitization of beta-adrenergic receptor-coupled adenylate cyclase in cerebral cortex after in vivo treatment of rats with desipramine. *J Neurochem* 47 (2): 454–459

Ordway GA (1997) Pathophysiology of the locus coeruleus in suicide. *Ann N Y Acad Sci* 836: 233–252

Ordway GA, Stockmeier CA, Cason GW, Klimek V (1997) Pharmacology and distribution of norepinephrine transporters in the human locus coeruleus and raphe nuclei. *J Neurosci* 17 (5): 1710–1719

Ortolano GA, Swonger AK, Kaiser EA, Hammond RP (1983) A calcium hypothesis of antidepressant action. *Med Hypoth* 10: 207–221

Ossowska G, Klenk-Majewska B, Szymczyk G (1997) The effect of NMDA antagonists on footshock-induced fighting behavior in chronically stressed rats. *J Physiol Pharmacol* 48 (1): 127–135

Oswald I, Brezinova V, Dunleavy DLF (1972) On the slowness of action of tricyclic antidepressant drugs. *Br J Psychiat* 120: 673–677

Papp M, Klimek V, Willner P (1994) Effects of imipramine on serotonergic and beta-adrenergic receptor binding in a realistic animal model of depression. *Psychopharmacology* (Berl) 114 (2): 309–314

Papp M, Moryl E (1994) Antidepressant activity of non-competitive and competitive NMDA receptor antagonists in a chronic mild stress model of depression. *Eur J Pharmacol* 263 (1–2): 1–7

Paul IA (1997) NMDA receptors and affective disorders. In: Antidepressants: new pharmacological strategies, Skolnick P (ed), Humana Press, Totowa, NJ, 145–158

Paul IA, Layer RT, Skolnick P, Nowak G (1993) Adaptation of the NMDA receptor in rat cortex following chronic electroconvulsive shock or imipramine. *Eur J Pharmacol* 247(3): 305–311

Paul IA, Nowak G, Layer RT, Popik P, Skolnick P (1994) Adaptation of the N-methyl-D-aspartate receptor complex following chronic antidepressant treatments. *J Pharmacol Exp Ther* 269 (1): 95–102

Paul IA, Trullas R, Skolnick P, Nowak G (1992) Down-regulation of cortical beta-adrenoceptors by chronic treatment with functional NMDA antagonists. *Psychopharmacology* (Berl) 106: 285–287

Pilc A, Legutko B (1995a) Antidepressant treatment influences cyclic AMP accumulation induced by excitatory amino acids in rat brain. *Pol J Pharmacol* 47 (4): 359–361

Pilc A, Legutko B (1995b) The influence of prolonged antidepressant treatment on the changes in cyclic AMP accumulation induced by excitatory amino acids in rat cerebral cortical slices. *Neuroreport* 7 (1): 85–88

Prikhozhan AV, Kovalev GI, Raevskii KS (1990) [Effects of antidepressive agents on glutamatergic autoregulatory presynaptic mechanism in the rat cerebral cortex]. *Biull Eksp Biol Med* 110 (12): 624–626

Przegalinski E, Tatarczynska E, Deren-Wesolek A, Chojnacka-Wojcik E (1997) Antidepressant-like effects of a partial agonist at strychnine- insensitive glycine receptors and a competitive NMDA receptor antagonist. *Neuropharmacology* 36 (1): 31–37

Pucilowski O (1997) Calcium channel antagonists in mood disorders. In: Antidepressants: new pharmacological strategies, Skolnick P (ed), Humana Press, Totowa, NJ, 81–102

Reynolds IJ, Miller RJ (1988) Tricyclic antidepressants block N-methyl-D-aspartate receptors: similarities to the action of zinc. *Br J Pharmacol* 95 (1): 95–102

Rosin DL, Melia K, Knorr AM, Nestler EJ, Roth RH, Duman RS (1995) Chronic imipramine administration alters the activity and phosphorylation state of tyrosine hydroxylase in dopaminergic regions of rat brain. *Neuropsychopharmacology* 12 (2): 113–121

Rossby SP, Sulser F (1997) Antidepressants: Beyond the synapse. In: Antidepressants: new pharmacological stratagies, Skolnick P (ed), Humana Press, Totowa, NJ, 195–212

Schildkraut JJ, Schanberg SM, Breese GR, Kopin IJ (1967) Norepinephrine metabolism and drugs used in the affective disorders: A possible mechanism of action. *Am J Psychiatry* 124: 600–608

Schildkraut JJ, Winokur A, Draskoczy PR, Hensle JH (1971) Changes in norepinephrine turnover in rat brain during chronic administration of imipramine and protriptyline: A possible explanation for the delay in onset of clinical antidepressant effects. *Am J Psychiatry* 127: 1032–1039

Shors T, Seib T, Levine S, Thompson R (1989) Inescapable versus escapable shock modulates long-term potentiation in the rat. *Science* 244: 224–226

Sills MA, Loo PS (1988) Tricyclic antidepressants and dextromethorphan bind with higher affinity to the phencyclidine receptor in the absence of magnesium and L-glutamate. *Mol Pharmacol* 36: 160–165

Silva MT, Rose S, Hindmarsh JG, Aislaitner G, Gorrod JW, Moore PK, Jenner P, Marsden CD (1995) Increased striatal dopamine efflux *in vivo* following inhibition of cerebral nitric oxide synthase by the novel monosodium salt of 7-nitro indazole. *Br J Pharmacol* 114 (2): 257–258

Sinei KA, Redfern PH (1993) The effects of chronic clomipramine and mianserin on the activity of tryptophan hydroxylase in the rat brain. *East Afr Med J* 70 (11): 721–724

Skolnick P (1999) Antidepressant for the new millennium. *Eur J Pharmacol* 375: 31–40

Skolnick P, Layer RT, Popik P, Nowak G, Paul IA, Trullas R (1996) Adaptation of N-methyl-D-aspartate (NMDA) receptors following antidepressant treatment: implications for the pharmacotherapy of depression. *Pharmacopsychiatry* 29 (1): 23–26

Skolnick P, Miller R, Young A, Boje K, Trullas R (1992) Chronic treatment with 1-aminocyclopropanecarboxylic acid desensitizes behavioral responses to compounds acting at the N-methyl-D-aspartate receptor complex. *Psychopharmacology* (Berl) 107 (4): 489–496

Southam E, Garthwaite J (1993) The Nitric Oxide-Cyclic GMP Signalling Pathway in Rat Brain. *Neuropharmacology* 32: 1267–1277

Stahl SM (1998) Basic psychopharmacology of antidepressants, part 1: Antidepressants have seven distinct mechanisms of action. *J Clin Psychiatry* 59 Suppl 4: 5–14

Stone EA (1983) Problems with the current catechholamine hypothesis of antidepressant agents: speculations leading to a new hypothesis. *Behav Brain Sci* 6: 555–577

Strasser A, McCarron RM, Ishii H, Stanimirovic D, Spatz M (1994) L-arginine induces dopamine release from the striatum *in vivo*. *Neuroreport* 5 (17): 2298–2300

Sugrue MF (1985) Delayed biochemical changes following antidepressant treatment. *Psychopharmacol Bull* 21: 619–622

Trullas R (1997) Functional NMDA antagonists: A new class of antidepressant agents. In: Antidepressants: new pharmacological strategies, Skolnick P (ed), Humana Press, Totowa, NJ, 103–124

Trullas R, Folio T, Young A, Miller R, Boje K, Skolnick P (1991) 1-aminocyclopropanecarboxylates exhibit antidepressant and anxiolytic actions in animal models. *Eur J Pharmacol* 203 (3): 379–385

Trullas R, Jackson B, Skolnick P (1989) 1-aminocyclopropanecarboxylic acid, a ligand of the strychnine-insensitive glycine receptor. *Pharmacol Biochem Behav* 34: 313–316

Trullas R, Skolnick P (1990) Functional antagonists at the NMDA receptor complex exhibit antidepressant actions. *Eur J Pharmacol* 185: 1–10

Trulson ME, Arasteh K, Ray DW (1986) Effects of elevated calcium on learned helplessness and brain serotonin. *Pharmacol Biochem Behav* 24: 445–448

Vetulani J (1991) The development of our understanding of the mechanism of action of antidepressant drugs. *Pol J Pharmacol Pharm* 43 (4): 323–338

Vetulani J, Stawarz RJ, Sulser F (1976) Adaptive mechanisms of the noradrenergic cyclic AMP generating system in the limbic forebrain of the rat: adaptation to persistent changes in the availability of norepinephrine (NE). *J Neurochem* 27 (3): 661–666

Vetulani J, Sulser F (1975) Action of various antidepressant treatments reduces reactivity of noradrenergic cyclic AMP-generating system in limbic forebrain. *Nature* 257 (5526): 495–496

Watkins CJ, Pei Q, Newberry NR (1998) Differential effects of electroconvulsive shock on the glutamate receptor mRNAs for NR2A, NR2B and mGluR5b [In Process Citation]. *Brain Res Mol Brain Res* 61 (1–2): 108–113

Wedzony K, Klimek V, Nowak G (1995) Rapid down-regulation of beta-adrenergic receptors evoked by combined forced swimming test and CGP 37849 – a competitive antagonist of NMDA receptors. *Pol J Pharmacol* 47 (6): 537–540

Wolfe BB, Harden TK, Sporn JR, Molinoff PB (1978) Presynaptic modulation of beta adrenergic receptors in rat cerebral cortex after treatment with antidepressants. *J Pharmacol Exp Ther* 207: 446–457

Yu PH, Boulton AA (1990) Effect of trimipramine, an atypical tricyclic antidepressant, on the activities of various enzymes involved in the metabolism of biogenic amines. *Prog Neuropsychopharmacol Biol Psychiatry* 14 (3): 409–416

Zhu MY, Blakely RD, Apparsundaram S, Ordway GA (1998) Down-regulation of the human norepinephrine transporter in intact 293- hNET cells exposed to desipramine. *J Neurochem* 70 (4): 1547–1555

Zhu MY, Ordway GA (1997) Down-regulation of norepinephrine transporters on PC12 cells by transporter inhibitors. *J Neurochem* 68 (1): 134–141

Zhu XZ, Luo LG (1992) Effect of nitroprusside (nitric oxide) on endogenous dopamine release from rat striatal slices. *J Neurochem* 59 (3): 932–935

Markers of depression

The hypothalamic-pituitary-adrenal axis and antidepressant action

Timothy G. Dinan

Department of Psychiatry, Royal College of Surgeons in Ireland, St. Stephen's Green, Dublin 2, Ireland

The hypothalamic-pituitary-adrenal axis (HPA) is the core stress axis in man, and together with the sympathoadrenal medullary system (SAM) it co-ordinates response to the diverse range of stressors, from psychological to physical. There is considerable interplay between both neuronal systems, especially between the noradrenergic nucleus locus coeruleus which provides central regulation of the SAM and the parvocellular neurones which regulate the HPA. The SAM by triggering catecholamine release provides the acute stress response, whilst the HPA governs longer term stress defence mechanisms. Together these systems regulate energy utilisation and metabolic activity throughout the body.

Corticotropin releasing hormone (CRH) is the principal central peptide responsible for stimulating ACTH release from the anterior pituitary. It is a 41 amino acid peptide originally discovered and sequenced by Vale and colleagues [1]. The main site of CRH production is the medial parvocellular neurones of the paraventricular nucleus of the hypothalamus. These neurones project to the external zone of the median eminence, where CRH is released into the portal vasculature, to act on CRH1 receptors on the anterior pituitary. It is now established that argenine vasopressin (AVP) acts as a co-secretogogue [2]. On its own AVP has little impact on ACTH secretion. However, in the presence of CRH, AVP synergistically interacts to produce a major increase in ACTH release. Many cells in the parvocellular neurones which under basal conditions produce CRH, co-express AVP in situations of stress. AVP produces its impact on the anterior pituitary through V1b receptors. The synergistic interaction between CRH and AVP is mediated at a second messenger level where the two receptor systems communicate.

A variety of neuronal inputs to the parvocellular neurones regulate CRH release, together with peripheral cytokines, which are produced as part of the acute phase immune response. Interleukin-1 and interleukin-6 are the most potent activators of the HPA.

ACTH is produced in the corticotroph cells of the anterior pituitary. It is a 39 amino acid peptide derived from the pro-hormone pro-opiomelanocortin (POMC) [3]. This peptide has its gene expression regulated by several factors including CRH. The pro-hormone also gives rise to β-lipotropin and β-endorphin. The ACTH receptor is a G-protein coupled receptor which uses cAMP as its second messenger system.

Cortisol is the principal glucocorticoid in man. Like all adrenocortical hormones its precursor is cholesterol which is taken up from the plasma by the adrenal cells. The synthesis of glucocorticoids occurs largely in the zona fasciculata with a small portion also produced in the zona reticularis. The release of cortisol undergoes significant diurnal variation, with high output in early morning and troughs occurring in the evening. The limited cortisol storage capacity of the adrenocortical cells requires a capacity for rapid activation of the synthetic sequence in situations of stress and especially the rate-limiting P450 desmolase reaction. Transcortin is a glycoprotein which binds cortisol and in the process renders it biologically inactive. Approximately 80% of cortisol is protein bound. The plasma half-life of cortisol is 70 min and the free fraction is filtered by the kidney.

The view of a central biological stress response was put forward by Selye in the 1930's as part of his theory of the general adaptation syndrome [4]. Work in adrenalectomised animals indicated that stressors which the animal would normally survive may prove fatal in the absence of glucocorticoids.

The precise manner in which glucocorticoids enable us to adapt to stress is far from understood. What is clear is that many brain regions, not just the hypothalamus, are an important target for peripherally produced glucocorticoids and that high levels of glucocorticoids are essential as an acute reaction to stress, while chronically elevated levels may have negative impacts on neuronal activity.

Glucocorticoid receptors
Our knowledge of the functional activity and pharmacology of central glucocorticoid receptors rests largely on studies conducted in rodents. Two types of central adrenal steroid receptor have been described, type 1 or mineralcorticoid receptors and type 2 or glucocorticoid receptors [5]. The type 1 receptors have a high affinity for corticosterone in the rat and are found in high concentrations in extrahypothalamic limbic neurones of the septo-hippocampal projection, whereas in the pituitary relatively low levels are found [6]. Overall the type 1 receptor is thought to mediate tonic influences of cortisol or corticosterone. The type 2 receptors are physicochemically indistinguishable from the peripheral glucocorticoid receptor and have high affinity for naturally occurring glucocorticoids. The highest concentration of these receptors in the brain is found in the dentate gyrus, lateral septum, paraventricular nucleus and in the brainstem nucleus tractus solitarius and nucleus locus coeruleus. When animals are adrenalectomised there is immediate upregulation of these receptors, which are downregulated by dexamethasone treatment.

Stress related cortisol changes are mediated by the type 2 receptor. The basal occupancy of the receptor is low and increases as cortisol levels rise in response to stress. At resting morning levels of plasma cortisol type 2 receptor occupancy is estimated to be around 10%. In stress occupancy in excess of 75% is possible.

HPA activity is regulated by a balance between forward drive and a series of negative feedback loops to the pituitary, hypothalamus and limbic structures. Negative feedback is described as immediate (within minutes), intermediate and delayed (over hours). The dexamethasone suppression test is the classic test of delayed feedback.

Depression and the HPA

Disturbances in HPA activity are the most widely documented biological alterations seen in major depression. Studies of HPA function in depressed patients reveal numerous abnormalities, which are more pronounced in patients with melancholic features [7]. The following abnormalities are well documented: (1) 24 h urinary free cortisol is enhanced (on average twice as high as in healthy controls) and serum cortisol levels are raised. (2) Dexamethasone non-suppression is found in 40–70% of patients. (3) Adrenal gland hyperplasia is reported. (4) Release of ACTH in response to CRH challenge is blunted.

Baseline HPA measures

Several studies have examined baseline levels of the key HPA hormones in patients with depression. Plasma CRH levels have been investigated by Catalan et al. [8]. Thirty-six patients with either major depression or dysthymia, together with 17 healthy volunteers were recruited. Plasma cortisol and CRH concentrations were significantly higher in the patient group. Plasma cortisol and CRH concentrations correlated significantly, suggesting the CRH is of hypothalamic origin.

CRH immunoreactivity in the CSF is also increased in depression. Nemeroff's group in a series of studies have found consistently elevated CRH, when depressives are compared with other diagnostic categories [9]. In contrast, patients with atypical features of depression such as hyperphagia, hypersomnia and rejection sensitivity have below normal levels of CRH [10].

Several studies have found elevated plasma cortisol levels in depression. Single plasma samples are unreliable in detecting alterations even in Cushing's disease. However multiple sampling over a short time-frame is reported as increasing the diagnostic reliability of cortisol in distinguishing depression from health. Halbreich et al. [11] monitored cortisol between 1300 and 1600 hours and claim a high sensitivity for this approach.

Salivary cortisol is a convenient painless method of assessing cortisol. As in the case of plasma and urine, salivary cortisol is reported as elevated in depression [12]. Depressed patients with an endogenous symptom complex had salivary cortisol and plasma ACTH measured. In healthy controls a strong correlation was established between plasma ACTH and salivary cortisol, but this relationship was lost in depression. Salivary cortisol was elevated in the depressives.

Baseline treatment effects
Treatment with fluoxetine decreases CRH levels in CSF [13] and also decreases levels of AVP. That elevated CRH is a state marker is also supported by a study in which the peptide was measured before and after a course of ECT [14]. Nine patients with major depression and psychotic features were recruited. CSF concentrations of CRF, somatostatin and beta-endorphin were measured. Concentrations of both CRF and beta-endorphin decreased after ECT, while the concentration of somatostatin increased.

In response to treatment with conventional antidepressants plasma cortisol levels decrease [15] and similar decreases in urinary free cortisol are reported [16].

Linkowski et al. [17] monitored ACTH and cortisol concentrations at 15 min intervals in 11 men suffering from major depression and a healthy comparison group. During the acute phase of illness the patients had abnormally short rapid eye movement latencies, hypercortisolism, early timing of the nadirs of the ACTH-cortisol rhythms and shorter nocturnal periods of quiescent cortisol secretion. They were treated either with electroconvulsive therapy or amitriptyline and after treatment cortisol levels returned to normal due to a decrease in the magnitude of episodic pulses. Furthermore, the timing of the circadian rhythms of ACTH and cortisol as well as the duration of the quiescent period of cortisol secretion was normalised. The authors conclude that a disorder of circadian rhythmicity characterises acute episodes of major depression and that this chronobiological abnormality as well as the hypersecretion of ACTH and cortisol are state rather than trait markers. Patients who fail to respond to treatment do not show such alterations.

Dynamic HPA Function Tests

Dexamethasone suppression test
The dexamethasone suppression test (DST) is a test of negative feedback inhibition within the HPA. It is the most extensively studied biological marker in psychiatry [18]. The standardised test consists of administering dexamethasone 1 mg at 11 p.m. and obtaining blood for cortisol estimation at 4 p.m. the following afternoon [19]. Non-suppression is usually defined as a cortisol level above 137 nmol/l (5 µg/dl). Most studies suggest the test has 40–50% sensitivity and 70–90% specificity in the diagnosis of major depression.

Treatment and the DST
A recent meta-analysis has focused on the predictive value of the test [20]. When studies in which a placebo was employed are examined non-suppressors were significantly less likely than suppressors to respond to placebo. Furthermore, suppressors responded better to antidepressants than placebo, indicating the test does not usefully select patients who will respond to pharmacotherapy. Therefore, the baseline DST result is poorly predictive of outcome. The test does provide a state dependent marker, as Carroll [21] first reported. Effective treatment should normalise the response and persistent non-suppression is associated with a poor outcome. This observation was supported by Goldberg [22] and Greden et al. [23] who found that persistent non-suppression, despite apparent clinical improvement, could serve as a marker of active underlying illness. Subsequent studies comparing patients whose DST normalises with those who do not, indicate that persistent non-suppression is associated with re-hospitalisation, suicide and recurrence of severe symptoms [24–26].

CRH test
The CRH test is conducted by the intravenous administration of either ovine or human CRH and the monitoring of ACTH and cortisol output. Because of differing protein binding characteristics the response to ovine CRH is longer than that observed with the human variety. Numerous studies consistently demonstrate that ACTH output is blunted but cortisol output is normal following CRH administration in depression [27–29].

Treatment and the CRH test
Amsterdam et al. [30] examined ovine CRH responses both before and during clinical recovery in depressed patients. Cumulative ACTH responses increased significantly during clinical recovery and were similar to those of normal subjects on recovery. Paradoxically, maximum and peak cortisol responses increased after recovery, suggesting that heightened adrenocortical responsiveness to ACTH during depression may take longer to normalise than abnormal pituitary responsiveness to CRH.

The test was found to be a good predictor of relapse in bipolar depression [31]. The ACTH and cortisol response to 100 µg of human CRH was measured in 42 lithium treated patients in remission from bipolar 1 disorder. Patients showed higher baseline and peak ACTH concentrations than healthy comparison subjects. A lower net area under the ACTH curve predicted depressive relapse within 6 months. The study suggests that the CRH test is potentially a good predictor of depressive relapse in remitted bipolar patients.

Combined dexamethasone/CRH test
When healthy volunteers are treated with dexamethasone prior to CRH infusion (dex/CRH test), release of ACTH is blunted and the extent of

blunting is proportional to the dose of dexamethasone. Paradoxically, when depressives are pretreated with dexamethasone they show enhanced response to CRH. The reason for this response is unclear but Holsboer [32] suggests that vasopressin may be involved. An alternative explanation is that dexamethasone results in a build-up of ACTH in the pituitary. As there is enhanced forward drive in depression greater levels of ACTH are available for release on CRH challenge.

Heuser et al. [33] estimates the sensitivity of the test to be 80% in differentiating a healthy subject from a depressive.

Treatment and dex/CRH test

The impact of amitriptyline on the dex/CRH test was assessed by Heuser et al. [34]. A total of 39 depressed inpatients, with a mean age of 69 years and a mean Hamilton depression score of 26 were recruited. The test was conducted at baseline and the end of weeks 1, 3 and 6 of treatment. Treatment consisted of amitriptyline 75 mg nocte.

Overall there was a continuous drop in the severity of depression throughout the 6 weeks of treatment. The patients' mean steady-state plasma concentration of tricyclic (amitriptyline+nortriptyline) was 180 ng/ml. At baseline the patients had a profoundly abnormal HPA response, characterised by an exaggerated cortisol release in response to dex/CRH. The abnormality began to disappear after 1 week of treatment.

Holsboer-Trachsler et al. [35] examined dex/CRH responses during 5 weeks of treatment in 42 major depressives given trimipramine or trimipramine plus sleep deprivation or bright light therapy. For the total sample, the treatment non-responders had significantly greater ACTH secretion following CRH at both the beginning and end of the study. The test is therefore viewed as having predictive value in determining response.

Serotonin and the HPA

A significant serotoninergic input to the hypothalamus regulates CRH release and, of the many 5HT receptors, the 1a receptor seems dominant in this response [36]. Stimulation of these receptors in humans is known to not only activate the HPA but also to induce hypothermia. Lesch [37] used ipsapirone, the azaspirone which acts as a partial agonist at the 5HT1a receptor, as a challenge in both depressives and healthy controls. Twenty-four subjects, 12 with unipolar depression and 12 matched healthy subjects were recruited. They were each given ipsapirone 0.3 mg/kg or placebo in random order. High basal cortisol levels were found in the depressives. The ACTH/cortisol and hypothermic responses were attenutated in patients with unipolar depression. The author suggests that the impaired HPA response in depression may be due to a glucocorticoid-dependent subsensitivity of the post-synaptic 5HT1a receptor or a defective post-receptor signaling pathway.

Lesch [37] examined the impact of amitriptyline treatment on 5HT1a induced hypothermia. Patients with major depression were chronically treated with amitriptyline and temperature response to ipsapirone challenge was monitored before and after treatment. Amitriptyline caused further impairment in 5HT1a mediated hypothermia. The study adds weight to the view that effective antidepressants down-regulate the 5HT1a receptor during treatment. A study of fluoxetine, using a double-blind placebo controlled design, in patients with obsessive compulsive disorder yielded similar findings [38]. The ability of ipsapirone (0.3 mg/kg) to induce hypothermia and ACTH/cortisol release was attenuated during fluoxetine challenge.

Synacthen tests
Thakore et al. [39] recruited female patients fulfilling criteria for major depression, melancholic subtype. Subjects were treated with a selective serotonin reuptake inhibitor, either sertraline (50 mg) or paroxetine (20 mg). Whilst depressed and after treatment, when again medication-free, patients underwent an ACTH test (tetracosactrin), 250 mg given as an intravenous bolus. Cortisol response to ACTH was measured as the level of cortisol post-ACTH relative to baseline. Treatment resulted in a significant drop in this value from 1633.3 ± 378.5 nmol/l (mean $\pm$ SEM) to 595.1 ± 207.7 nmol/l. Successful pharmacological treatment of depression appears to be associated with a reduction in ACTH-induced cortisol release in drug-free patients.

When baseline HPA characteristics of patients who relapse are compared with those who do not, significant differences are observed [40]. A group of patients who required continuing treatment to hold their symptoms in abeyance, were compared with patients who were successfully tapered off medication without remission. Relapsers had high baseline ACTH and cortisol together with enhanced cortisol response to ACTH. The authors suggest that abnormal baseline HPA measures increase the necessity for continuation treatment of depression.

Adrenal gland volume and antidepressants
Adrenal gland volume was measured using magnetic resonance imaging in 11 patients (nine adults and two adolescents) with major depression [41]. All subjects were assessed during the depressive episode and in full remission, when they were drug-free for at least 1 month. A group of healthy age- and sex-matched comparison subjects were also tested. Baseline ACTH and cortisol was measured, together with CRH and ACTH responses.

Mean adrenal gland volume was approximately 70% larger in the depressives than healthy comparison subjects and reduced on average by 70% after treatment. Post-treatment differences in adrenal volume were no longer present. Patients and healthy subjects had adrenal glands of similar volume. Baseline hormone measurements and responses to CRH or ACTH did not differ significantly between the two groups. The authors conclude that adrenal gland enlargement occurring during an episode of major depression is state-dependent, in that it reverts to the normal size range during remission.

HPA and antidepressants

It is generally assumed that standard antidepressants act by blocking the uptake of monoamines and consequently inducing receptor alterations. However, they also exert important influences on glucocorticoid receptors. The effect of long term imipramine treatment on glucocorticoid receptor immunoreactivity in various regions of the rat brain has been investigated [42]. A significant upregulation of corticoid receptors was found on the noradrenergic nucleus locus coeruleus and the serotonin containing raphe neurones. A similar upregulation occurs in the hypothalamus and hippocampus and thereby induces an increase in negative feedback, decreasing the overall acitivty of the HPA. These finding have been replicated in *in situ* hybridisation studies and the results are found to be applicable to the selective serotonin reuptake inhibitors [43].

This action of antidepressants may account for their capacity to suppress HPA activity and may be of clinical relevance. Recent papers from the Max Planck Institute in Munich suggest that the HPA may be an appropriate target site for pharmacological intervention in depression [44]. If this is the case, which aspects of HPA function may provide a suitable avenue for pharmacological manipulation in the management of depression?

The most extensively investigated site within the HPA for pharmacological intervention in the treatment of depression has been the adrenal cortex. Several investigators have made use of glucocorticoid synthesis inhibitors as potential antidepressants. Three synthesis inhibitors have been used in studies to date: (1) metyrapone, an 11β-hydroxylase inhibitor; (2) aminoglutethimide, which blocks the conversion of cholesterol to pregnenolone; (3) ketoconazole which inhibits cortisol synthesis acting at several enzyme sites and also acting as a type 2 receptor antagonist.

Murphy and colleagues [45] provided the most extensive investigation of these agents. They have reported on 20 patients with treatment resistant depression who were treated with either aminoglutethimide 250–1000 mg/day, ketoconazole 400–1200 mg/day or metyrapone 250–1000 mg/day. Most patients were treated for a total of 8 weeks. A response rate of 50% was reported. Thakore and Dinan [46] used ketoconazole 400–600 mg/day for 4 weeks in 8 melancholically depressed inpatients. Serotoninergic responses were measured before and after treatment using the d-fenfluramine/prolactin stimulation test. Five of the eight patients showed an excellent response and three had a partial response to treatment. Clinical improvements paralleled the decrease in cortisol levels and an enhancement in serotoninergic responsivity.

At the Maudsley Hospital, O'Dwyer et al. [47] conducted a single-blind study in depressed inpatients who received either metyrapone 1000 mg to 2000 mg/day together with hydrocortisone or placebo for 2 weeks. A crossover design was employed.

The authors conclude that metyrapone and hydrocortisone treatment resulted in an alleviation of depression, as evidenced by a decrease in Hamilton depression scale scores. Wolkowitz et al. [48] used ketoconazole 400–800 mg/day for 3–6 weeks in 10 patients with major depression. A decrease of approximately 30% in Hamilton scores were observed. Less promising results were obtained by Amsterdam and Hornig-Kernig [49] who found no response in 10 patients with refractory depression treated with ketoconazole.

Whether or not this strategy can yield a treatment for depression with benefits in excess of those provided by current antidepressants will not be answered by further small scale studies. A full placebo-controlled study in an adequate sample size is necessary.

Mineralocorticoid and Glucocorticoid receptors

Two further sites for pharmacological intervention are the mineralocorticoid and glucocorticoid receptors. It has been argued that antagonists at these receptor sites may have antidepressant properties. Mifepristone (RU 486) is not only a potent antagonist of sex steroid receptors, but also of glucocorticoid receptors. Related compounds with greater specificity for glucocorticoid receptors (e.g. Org 34517) are now available and may have therapeutic impact in depression. Only preliminary observations suggesting efficacy have so far been published.

CRH and AVP antagonists

Non-peptide CRH and AVP antagonists are now available and may be of benefit in adrenal hypersecretory states such as melancholic depression. Preliminary work with antalarmin, a novel pyrrolopyrimidine compound, suggests potential as an antidepressant. It acts as a CRH1 receptor antagonist, displacing CRH from binding to the receptor in the anterior pituitary, frontal cortex and cerebellum. It potently inhibits CRH-stimulated ACTH release. It seems likely that CRH antagonists will enter full-scale clinical trials shortly.

In depression, recent evidence suggests the CRH1 receptor on the anterior pituitary downregulates. If this is the case is it likely that a CRH antagonist would rectify the HPA disturbance observed in the disorder? A study by Dinan et al. [50] indicates that AVP receptor blockade may have more potential. Patients with major depression underwent testing on two separate occasions. One test consisted of CRH stimulation and the second of CRH plus ddAVP stimulation. A significant blunting of ACTH output in response to CRH challenge was noted. When patients were given both CRH and ddAVP the release of ACTH in depressives and healthy volunteers was indistinguishable. These findings suggest that whilst the CRH1 receptor is downregulated in depression, a concomitant upregulation of the V1b receptor takes place. This is consistent with animal models of chronic stress where a switching from CRH to AVP regulation of the HPA is observed. Increased production of AVP leads to an upregulation of the V1b receptor. The results suggest that a blockade of the V1b receptor offers the most appropriate pharmacological strategy for normalising the endocrine component of the depressive syndrome.

Conclusions

In the development of new antidepressants the HPA offers several possible target sites. Since the development of the first antidepressants in the 1950s pharmaceutical companies have targeted monoamine systems in a search for new compounds.

Whilst this approach has now produced a generation of compounds relatively safe in overdoses and well tolerated, it has not yet produced compounds that act rapidly and are effective in more than 70% of cases. A change in target site from monoamines to HPA may yield such benefits. The results of placebo-controlled studies of agents that act directly on the HPA are awaited with interest.

References

1 Vale W, Spiess J, Rivierc et al (1981) Characterisation of a 41 residue ovine hypothalamic peptide that stimulates secretion of the corticotropin and beta-endorphin. *Science* 213: 1394–1399
2 Scott LV, Dinan TG (1998) Vasopressin and the regulation of hypothalamic-pituitary-adrenal axis function: Implications for the pathophysiology of depression. *Life Sciences* 62: 1985–1998
3 Axelrod J, Reisine TD (1984) Stress hormones: their interaction and regulation. *Science* 224: 452–459
4 Selye H (1956) *The stress of life*. McGraw-Hill, New York
5 McEwen BS, Davis PG, Parsons B (1979) The brain as a target for steroid hormone action. *Ann Rev Neurosci* 2: 65–112
6 Reul JMHM, DeKloet R (1986) Anatomical resolution of two types of corticosterone receptor sites in rat brain with *in vitro* autoradiography and computerised image analysis. *J Steroid Biochem* 24: 296–304
7 Dinan TD (1994) Glucocorticoids and the genesis of depressive illness: A psychobiological model. *Brit J Psych* 164: 365–371

8. Catalan R, Gallart JM, Castellanos JM, Galard R (1998) Plasma corticotropin-releasing factor in depressive disorders. *Biol Psychiatry* 44: 15–20
9. Nemeroff CB, Owens MJ, Bissette G et al (1984) Elevated concentrations of CSF corticotropin-releasing factor-like immunoreactivity in depressed patients. *Science* 226: 1342–1344
10. Geracioti TD, Orth DN, Ekhator NN (1992) Serial cerebrospinal fluid corticotropin-releasing hormone concentration in healthy and depressed humans. *J Clin Endocrin Metab* 74: 1325–1330
11. Halbreich U, Zumoff B, Kream J, Fukushima DK (1982) The mean 1300–1600 h plasma cortisol concentration as a diagnostic test for hypercortisolism. *J Clin Endocrin Metab* 54: 1262–1264
12. Galard R, Gallart JM, Catalan R et al (1991) Salivary cortisol levels and their correlation with plasma ACTH levels in depressed patients before and after the DST. *Am J Psych* 148: 505–508
13. De Bellis MD, Gold PW, Geracoti D, Listwak S, Kling MA (1993) Fluoxetine significantly reduces CFS CRH and AVP concentrations in patients with major depression. *Am J Psych* 150: 656–657
14. Nemeroff CB, Bissette G, Akil H, Fink M (1991) Neuropeptide concentrations in cerebrospinal fluid of depressed patients treated with electroconvulsive therapy: Corticotropin releasing factor, β-endorphin and somatostadin. *Brit J Psych* 158: 59–63
15. O'Keane V, McLoughlin D, Dinan TD (1992) D-fenfluramine-induced prolactin/cortisol release in major depression: Response to treatment. *J Affect Disorders* 26: 143–150
16. Carroll BJ, Curtis GC, Davies BM et al (1976) Urinary free cortisol excretion in depression. *Psychol Med* 6: 43–51
17. Linkowski P, Mendelwicz J, Kerkhofs M et al (1987) 24 hour profiles of adrenocorticotropin, cortisol and growth hormone in major depression: effect antidepressant treatment. *J Clin Endocrin Metab* 65: 141–152
18. American Psychiatric Association Task Force (1987) The dexamethasone suppression test: An overview of its current state in Psychiatry. *Am J Psych* 144: 1253–1262
19. Carroll BJ (1982) The dexamethasone suppression test for melancholia. *Brit J Psych* 140: 292–304
20. Ribeiro SCM, Tandon R, Grunhaus L, Greden JF (1993) The DST as a predictor of outcome in depression: A meta-analysis. *Am J Psych* 150: 1618–1629
21. Carroll BJ (1968) Pituitary-adrenal function in depression. *Lancet* 1: 1373
22. Goldberg IK (1980) Dexamethasone suppression tests in depression and response to treatment. *Lancet* 2: 92
23. Greden JF, Albala AA, Haskett RF et al (1980) Normalisation of dexamethosone suppression test: A laboratory index of recovery from endogenous depression. *Bio Psychiatry* 15: 449–458
24. Holsboer F, Liebl R, Hofschuster E (1982) Repeated dexamethasone suppression tests during depressive illness. *J Affective Dis* 4: 93–101
25. Asnis GM, Halbreich U, Rabinowitz H et al (1986) The dexamethasone suppression test (1 mg and 2 mg) in major depression: illness versus recovery. *J Clin Psychopharma* 6: 294–296
26. Charles GA, Schittecatte M, Rush AJ et al (1989) Persistent cortisol nonsuppression after clinical recovery predicts symptomatic relapse in unipolar depression. *J Affective Dis* 17: 271–278
27. Holsboer F, Gerken A, Von Bardeleben U et al (1986) Human corticotropin-releasing hormone in depression. *Biol Psychiatry* 21: 609–611
28. Amsterdam JD, Maislin G, Winokur A, Kling M, Gold P (1987) Pituitary and adrenocorical responses to the ovine corticotropin-releasing hormone in depressed patients and healthy volunteers. *Arch Gen Psychiatry* 44: 775–781
29. Ur E, Dinan TG, O'Keane V et al (1992) Effect of metyrapone on the pituitary-adrenal axis and depression: Relation to dexamethasone suppressors status. *Neuroendocrin* 56: 533–539
30. Amsterdam JD, Maisling, Winokur et al (1988) The CRH stimulation test before and after clinical recovery from depression. *J Affective Dis* 14: 213–222
31. Vieta E, Gasto C, De Osaba MJM et al (1997) Prediction of depressive relapse in remitted bipolar patients using corticotropin-releasing hormone challenge test. *ACTA Psychiatrica Scandinavica* 95: 205–211

32 Holsboer FH (1995) Neuroendocrinology of mood disorders. In: FE Bloom, DJ Kuffer (eds): *Psychopharmacology: The Fourth Generation of Progress.* Raven Press, New York 957–970
33 Heuser I, Yassouridisa, Holsboer F (1994) The combined dexamethasone: A refined laboratory test for psychiatric disorders. *J Psychiatric Res* 28: 341–356
34 Heuser IJE, Schweiger U, Gotthard DT et al (1996) Pituitary-adrenal-system regulation and psychopathology during amitriptyline treatment in elderly depressed patients and normal comparison subjects. *Am J Psych* 153: 93–99
35 Holsboer-Trachsler E, Stohler R, Hatzinger M (1991) Repeated administration of the combined dexamethasone-human corticotropin releasinghormone stimulation test during treatment of depression. *Psychiat Res* 38: 163–171
36 Dinan TG (1997) Serotonin and the regulation of hypothalamic-pituitary-adrenal axis function. *Life Sciences* 58: 1683–1693
37 Lesch KP (1991) 5-HT1A receptor responsivity in anxiety disorders and depression. *Progress Neuropsychopharma Bio Psychiatry* 15: 723–733
38 Lesch KP, Hoh A, Schulte HM (1991) Long term fluoxetine treatment decreases 5-HT1A receptor responsivity in obsessive-compulsive disorder. *Psychopharmacology* 105: 415–420
39 Thakore JH, Barnes C, Joyce J, Dinan TG (1997) The effects of antidepressant treatment on corticotropin-induced cortisol responses in patients with melancholic depression. *Psychiatry Res* 73: 27–32
40 O'Toole SM, Seckula LK, Rubin RT (1997) Pituitary-adrenal cortical axis measures as predictors of sustained remission in major depression. *Biol Psychiatry* 42: 85–89
41 Rubin RT, Phillips JJ, Sadow TF, McCracken JT (1995) Adrenal gland volume in major depression: Increased during the depressive episode and decreased with successful treatment. *Arch Gen Psychiatry* 52: 213–218
42 Kitayama I, Janson AM, Cintra A (1988) Effects of chronic imipramine treatment on glucocorticoid receptor immunoreactivity in various regions of the rat brain. *J Neural Trans* 73: 191–203
43 Brady LS, Whitfield HJ, Fox R (1991) Longterm antidepressant administration alters corticotropin-releasing hormone, tyrosine hydroxylase and mineralcorticoid receptor gene expression in the rat. *J Clin Investig* 87: 831–837
44 Holsboer R, Barden N (1996) Antidepressants and hypothalamic-pituitary- adrenocortical regulation. *Endocr Rev* 17: 187–199
45 Murphy BE (1997) Antiglucocorticoid therapies in major depression: a review. *Psychoneuroendocrinol* 22: S125–145
46 Thakore JH, Dinan TG (1995) Cortisol synthesis inhibition: a new strategy for the clinical and endocrine manifestations of depression. *Biol Psychiat* 37: 364–368
47 O'Dwyer AM, Lightman SL, Marks MN, Checkley S (1995) Treatment of major depression with metyrapone and hydrocortisone. *J Affect Dis* 33: 123–128
48 Wolkowitz OM, Reus VI, Manfredi F (1993) Ketoconazonazole administration in hypercortisolemic depression. *Am J Psych* 150: 810–812
49 Amsterdam SD, Hornig-Rohan M (1993) Adrenocortical activation and steroid suppression with ketoconazole in refractory depression. *Biol Psychiat* 33: 88A
50 Dinan TG, Lavelle E, Scott LV, Newell-Price J, Grossman A (1999) Desmopressin normalises the blunted ACTH response to corticotropin releasing hormone in melancholic depression: evidence of enhanced vasopressinergic responsivity. *J Clinendocrinol Metab* 84: 2238–2246

Neuroendocrine markers of depression and antidepressant drug action

Philip J. Cowen

University Department of psychiatry, Warneford Hospital, Oxford OX3 7JX, UK

Introduction

Investigating the role of neurotransmitters in central nervous system disorders is constrained by the difficulty of studying the neuropharmacology of the living human brain directly. Recent technological advances, however, are increasing the possibilities in this area. For example, the use of appropriately labelled ligands in conjunction with positron emission tomography (PET) and single photon emission tomography (SPET) allows the assessment of the density of specific neurotransmitter binding sites in discrete brain regions [1]. Furthermore the use of ligand binding together with pharmacological manipulation of endogenous neurotransmitter release can be used to measure neurotransmitter release *in vivo* [2].

While these advances open exciting possibilities there is still a need for relatively non-invasive tests that can be used to measure dynamic aspects of neurotransmitter function *in vivo*. Neuroendocrine challenge tests offer a simple and feasible means of achieving this goal.

The use of neuroendocrine challenge tests rests on the fact that the hypothalamus is heavily innervated by a diverse group of neurotransmitter pathways that regulate the secretion of peptide releasing factors into the portal circulation of the anterior pituitary gland and thus the secretion of certain pituitary hormones into plasma. Perhaps because of the richness of hypothalamic innervation and its complexity, resting or baseline levels of hormones rarely give useful information about the activity of specific neurotransmitters. However activation of a specific neurotransmitter by a selective pharmacological agent can lead to reliable and reproducible changes in plasma hormone levels. In this case it can be demonstrated that the size of the hormone response does indeed provide information about the functional consequences of stimulating the neurotransmitter pathway concerned [3].

The main requirement in such studies is a selective pharmacological probe that activates the neurotransmitter to be studied. Broadly, such probes work either through increasing the pre-synaptic activity of the neurotransmitter or by stimulating pre or post-synaptic receptors directly. Selection of

Table 1. Neuroendocrine measures of NA and DA function

Drug	Hormone Response	Mechanism
desipramine	↑ GH	indirect activation α_2-receptors
desipramine	↑ ACTH/cortisol	indirect activation α_1-receptors
clonidine	↑ GH	α_2-receptor agonist
apomorphine	↑ GH	dopamine D_2 receptor agonist
physiological	↑ melatonin	indirect activation β_1-receptors

a probe therefore depends on the hypothesis being tested and, of course, on the availability of an appropriate drug.

As well as changes in plasma hormone levels, selective drug challenge can produce other functional changes which can be measured. For example, selective serotonin (5-HT) receptor agonists can alter body temperature [4] and the α_2-adrenoceptor antagonist, clonidine causes sedation [5]. The value of such measurements is that they can provide information about aspects of neurotransmitter function brain regions other than those concerned in neuroendocrine regulation. They can therefore supplement the assessment of neurotransmitter function produced by hormonal measures.

The present chapter will review the more selective neuroendocrine probes that have been used to assess neurotransmitter function in depression (Tab. 1). Such probes can also be used to investigate how antidepressant drugs may alter brain neurotransmitter function. For the latter studies, animal experimental studies can provide a useful guide as to what might be expected in humans. Often in studies of antidepressant drug action the focus is on the effects of repeated antidepressant administration because the neuropharmacological consequences of repeated treatment are believed to be particularly relevant to the therapeutic properties of antidepressant drugs treatment.

This review will concentrate on neuroendocrine markers of catecholamine and 5-HT function, because most human work has been carried out on these systems. It is worth noting, however, that neuroendocrine abnormalities have been detected in other neurotransmitter systems in major depression, for example, in the growth hormone response to acetylcholine [6].

Noradrenaline

Pharmacological challenges

Many antidepressant drugs block the re-uptake of noradrenaline (NA) and drugs such as desipramine and maprotiline are relatively selective NA re-uptake inhibitors. Acute administration of desipramine increases plasma growth hormone (GH) and corticotropin (ACTH) and cortisol. Studies with selective receptor antagonists suggest that the GH response to desipramine is mediated *via* indirect activation of post-synaptic α_2-adrenoceptors while the ACTH and cortisol responses are produced by indirect activation of post-synaptic α_1-adrenoceptors [7, 8].

Clonidine is an α_2-adrenoceptor agonist that increases plasma GH levels. Its ability to stimulate α_2-adrenoceptors in other brain regions produces hypotension and sedation. The hypotensive effects appear to be mediated *via* activation of post-synaptic α_2-adrenoceptors but the location of the α_2-adrenoceptors involved in sedation are unclear, although a pre-synaptic location has been suggested [5]. Yohimbine is an α_2-adrenoceptor antagonist though it is not particularly selective. Acute administration of yohimbine increases cortisol and blood pressure [5].

The secretion of melatonin by the pineal gland has also been used as a marker of NA function. While it is difficult to elevate plasma melatonin in the day time with an acute pharmacological challenge, the night time increase in plasma melatonin reflects overall neurotransmission through pineal NA synapses. This involves activation of post-synaptic β_1-adrenoceptors by NA released from sympathetic nerve terminals innervating pinealocytes [9]. It is important to note that the pineal is innervated by the peripheral sympathetic nerve system. However, it is a useful model of central NA synapses in that the effect of repeated antidepressant treatment on pineal β_1-adrenoceptors appears essentially similar to the effects a central β_1-adrenoceptors [10].

NA neuroendocrine function in depressed patients

Earlier studies found reasonably consistent evidence that patients with major depression, particularly the melancholic subtype had blunted GH responses to clonidine and desipramine [5, 11]. More recent investigations, however, have often failed to find clear differences between depressives and controls, particularly when using lower doses of clonidine [12, 13]. Overall, however, GH responses to clonidine probably are blunted in depressed patients, particularly those who have been hospitalised and suffer more severe illnesses [14, 15].

As described below previous treatment with antidepressant drugs may cause a long-lasting suppression of GH responses to clonidine and presumably this may account for some or all of the blunting seen in depressed patients [16]. With this caveat the neuroendocrine data suggest that depressed patients may have a decreased sensitivity of certain post-synaptic α_2-adrenoceptors in depression. Depressed subjects, do not, however, seem to manifest differences in the hypotensive effects of clonidine which suggest that abnormalities in post-synaptic α_2-adrenoceptor function may be restricted to certain brain regions, notably the hypothalamus [5].

The effects of yohimbine in depression have been less studied but it has been reported that the cortisol and blood pressure responses are enhanced in depressed patients [17]. It is not easy to know what to make of this because the consequences of receptor down-regulation on the pharmacological effects of an antagonist are difficult to interpret. In addition, depressed patients often exhibit baseline elevations of cortisol which could conceivably interfere with pharmacological challenges that themselves use changes in plasma cortisol as an end-point.

If the GH responses to desipramine and clonidine are blunted in depressed patients it is important to ask whether this abnormality reflects changes in NA neurotransmission or rather intrinsic changes in GH secretion. It is difficult to answer this question definitively but in general GH responses to Growth Hormone Releasing Hormone (which acts directly at pituitary level) are not consistently decreased in depression [18, 19]. Accordingly it seems likely that the blunted GH responses to NA activation does reflect lowered sensitivity of post-synaptic NA receptors. Follow-up studies suggest that the decreased responsivity of hypothalamic α_2-adrenoceptors seen in depression is probably a trait marker, that is it persists after resolution of the depressed state [14, 20].

As noted above the secretion of the pineal hormone, melatonin, provides a marker of neurotransmission at β-adrenergic synapses. Studies of melatonin secretion in depressed patients have been somewhat contradictory with early investigations finding decreases but later and better controlled investigations finding, if anything a trend to increased overnight secretion [21, 22]. There is no clear evidence for a phase-shift in the secretion of melatonin in depression [21, 22].

Antidepressant treatment and NA neuroendocrine function

In animal experiments there is evidence that repeated treatment with tricyclic antidepressants (TCAs) can desensitise pre and post-synaptic α_2-adrenoceptors [23, 24]. Similar changes appear to occur in humans in that administration of both TCAs and the selective serotonin re-uptake inhibitor (SSRI) fluoxetine lower the GH response to clonidine and desipramine respectively [25, 26]. The blunting after tricyclic treatment appears to last at least 3 weeks. This suggests that TCA and probably SSRI treatment lowers the sensitivity of post-synaptic α_2-adrenoceptors.

Treatment with TCAs also appears to decrease the sedative and hypotensive responses to clonidine [23]. The former response may be mediated in part *via* activation of pre-synaptic α_2-adrenoceptors, which suggests that TCAs may decrease the sensitivity of this inhibitory autoreceptor [24]. This effect would be expected to enhance NA release and it is of interest that the primary pharmacological action of some antidepressants, notably, mirtazepine, is believed to be α_2-adrenoceptor blockade [27].

The effects of repeated antidepressant TCA treatment on melatonin secretion appear to differ between depressed patients and healthy controls. In healthy subjects repeated administration of TCAs initially increases melatonin secretion but this effect then attenuates until after about 3 weeks' treatment, melatonin levels have returned to baseline [28, 29]. It has been hypothesised that this decline in melatonin secretion represents a down-regulation of post-synaptic β_1-adrenoceptors which develops as a homeostatic response to increased pre-synaptic NA caused by the TCA. In depressed patients, however, TCA treatment leads to a sustained increase in night-time melatonin, indicating a persistent facilitation of β-adrenergic neurotransmission [28, 29]. The effects of SSRIs on melatonin secretion are rather contradictory. Although fluvoxamine has been reported to increase overnight melatonin secretion [30], fluoxetine treatment slightly decreased melatonin levels in both healthy subjects and patients with seasonal affective disorder [31].

Dopamine

Neuroendocrine challenges

Drugs that increase dopamine function increase plasma GH and lower prolactin (PRL). This effect of dopamine to inhibit PRL secretion is mediated *via* a direct effect at pituitary level while the GH responses probably reflects activation of DA receptors in the hypothalamus [32]. Both dopamine receptor agonists such as apomorphine and antagonists such a metoclopramide have been used in challenge studies in depressed patients.

DA neuroendocrine function in depressed patients

While early studies suggested that the GH response to apomorphine was normal in depressed patients more recent investigations have found a blunted response [33, 34]. This is consistent with impaired sensitivity of hypothalamic post-synaptic D_2 receptors. It is possible that previous antidepressant treatment could contribute to this effect (see below). However, Pichot et al. found blunted GH responses to apomorphine in depressed inpatients who had never received antidepressant medication [35]. There is no evidence that the decrease in PRL produced by dopamine agonists is altered in depressed patients [36]. A single study with the dopamine receptor antagonist, metoclopramide, found a trend to increased prolactin responses in depressed subjects [37]. These findings suggest that the senstivity of pituitary D_2 receptors is unaltered in depression.

Antidepressant treatment and DA neuroendocrine function

Animal studies suggest that repeated administration of many different kinds of antidepressant treatment increases the sensitivity of dopamine receptors in the mesolimbic forebrain [38]. However, there are few studies of the effects of antidepressant treatment on dopamine neuroendocrine function in humans. Administration of amitriptyline to both healthy subjects and depressed patients lowers the GH response to apomorphine [39] but this is probably attributable to the anticholinergic effects of amitriptyline.

Serotonin (5-HT)

Neuroendocrine challenges

Serotonin (5-HT) function has been widely studied in neuroendocrine work, firstly because of the indoleamine hypothesis that links low brain neurotransmission 5-HT with depressive disorder and secondly, because of the availability of selective 5-HT re-uptake inhibitor (SSRI) antidepressants [40]. Pre-synaptic 5-HT probes include the precursor, L-tryptophan (TRP) and the 5-HT releasing agent, fenfluramine which has been withdrawn because of cardiac toxicity. Intravenous antidepressants with high affinity for the 5-HT re-uptake site such as clomipramine and more recently, citalopram, have also been used. Most of these agents produce increases in plasma PRL and cortisol and TRP also increases plasma GH [4, 40].

Studies with selective receptor antagonists suggest that the PRL and GH responses to TRP are mediated *via* indirect activation of post-synaptic 5-HT_{1A} receptors. In contrast the PRL responses to fenfluramine appears to be a consequence of indirect activation of post-synaptic 5-HT_{2C} receptors [4, 40].

A number of direct 5-HT receptor agonists are also available for clinical study (Tab. 2). Most are 5-HT_{1A} receptor agonists and their administration to humans produces a characteristic profile of endocrine and temperature effects. The most reliable changes are an increase in plasma GH and a decrease in body temperature. Given in sufficient doses, most 5-HT_{1A} receptor agonists also increase plasma ACTH and cortisol. Some currently employed 5-HT_{1A} receptor agonists, notably buspirone and flesinoxan, also

Table 2. Neuroendocrine profile of 5-HT receptor agonists

Drug	Receptor	Responses
Ipsapirone	5-HT_{1A}	↑ ACTH, GH, ↓ TEMP
mCPP	5-HT_{2C}	↑ ACTH, PRL, ↑ TEMP
sumatriptan	$5\text{-HT}_{1B/1D}$	↑ GH, ↓ PRL

increase plasma PRL levels. However, the role of 5-HT$_{1A}$ receptors in this response is unclear [40–42].

Studies in animals suggest that endocrine responses to 5-HT$_{1A}$ receptor challenge are mediated by activation of post-synaptic 5-HT$_{1A}$ receptors. The hypothermic response appears to be a consequence of activation of cell body 5-HT$_{1A}$ autoreceptors but in some species post-synaptic 5HT$_{1A}$ receptors may be involved as well [4, 43].

Fewer studies have been reported of other 5-HT receptor challenges in depressed patients. M-chlorophenylpiperazine (mCPP) is a metabolite of the antidepressant drug, trazodone, and produces increases in plasma PRL, cortisol and body temperature. This is probably mediated *via* activation of 5-HT$_{2C}$ receptors [4, 45]. The group of "triptan" drugs, such a sumatriptan used to treat migraine, have a high affinity for 5-HT$_{1D}$ receptors. Acute administration of these compounds increases plasma GH and this is likely to be mediated *via* direct activation of post-synaptic 5-HT$_{1D}$ receptors [46].

5-HT neuroendocrine function in depressed patients

Five studies of drug-free depressed patients have reported blunted PRL and GH responses to intravenous TRP compared to healthy controls. Furthermore, upon clinical recovery the endocrine responses return to normal suggesting that the impaired endocrine responses are a state marker of depression. Interestingly the PRL response to the 5-HT re-uptake inhibitor, clomipramine, also appears to be consistently blunted in depression [4, 47].

Blunted PRL responses to fenfluramine have been found in about half the studies of drug-free depressed patients, so this abnormality does not seem to be as consistent as the finding of impaired PRL response to TRP [48]. This may be because fenfluramine acts indirectly to stimulate 5-HT$_{2C}$ rather than 5-HT$_{1A}$ receptors [4]. It also appears that blunted PRL responses to fenfluramine are associated with particular patient characteristics although studies implicate two rather different patient populations. The first tend to be inpatients, have severe depressive symptoms with melancholia and may demonstrate cortisol hypersecretion. The second have aggressive and impulsive personality traits with a history of suicide attempts [48, 49].

It is worthwhile noting that subjects in the latter group may manifest blunted PRL responses to fenfluramine in the absence of a current depressive disorder. Presumably, here the blunted PRL responses may represent a trait marker of 5-HT dysfunction and could, perhaps, correspond with the low levels of CSF 5-HIAA that have been reported in subjects who tend to behave in an aggressive and impulsive way [49].

Overall, therefore, the data suggest that depression is associated with decreased 5-HT-mediated PRL release. Importantly PRL responses to other pharmacological challenges such as the dopamine D$_2$ receptor antagonist, metoclopramide and the direct lactotroph stimulant, thyrotropin releasing hormone, are not reliably blunted in depressed patients [4, 37]. This suggests that the impairment in 5-HT-mediated PRL release seen in depression is not attributable to a generalised decrease in PRL secretion.

The investigations outlined above demonstrate that endocrine responses to drugs that activate pre-synaptic 5-HT neurones are likely to be impaired in major depression. This could reflect impaired 5-HT synthesis and release or lowered sensitivity of post-synaptic 5-HT receptors. The use of 5-HT neuro-endocrine probes that directly activate post-synaptic 5-HT receptors allows this question to be addressed but the findings are somewhat contradictory.

Two studies in drug-free depressed patients, one with buspirone and with ipsapirone, have found blunted hypothermic responses which could be consistent with lowered sensitivity of 5-HT_{1A} autoreceptors [41, 50]. In addition, studies that employed ipsapirone as a challenge found impaired cortisol release [41, 51]. While this is consistent with subsensitivity of post-synaptic 5-HT_{1A} receptors, the GH response to buspirone, another probable measure of post-synaptic 5-HT_{1A} receptor sensitivity, was unchanged in depressed patients [50]. It must be acknowledged, however, that ipsapirone is a better probe of 5-HT_{1A} receptors than buspirone because the latter drug also has primary actions on dopamine pathways.

Endocrine responses to other post-synaptic receptor agonists have been less studied in drug-free depressed patients. However PRL and cortisol responses to mCPP appear to be unchanged [4, 47]. All three studies that examined the GH response to sumatriptan in drug-free depressed patients found blunted responses compared to healthy controls [46]. This suggests a lowered sensitivity of the 5-HT_{1D} receptors involved in mediating GH release. These receptors are presumably located post-synaptically [42].

The studies outlined above demonstrate that there is good evidence that certain aspects of brain 5-HT function are impaired in major depression. At present the most consistent abnormalities are found with drugs that act pre-synaptically to increase brain 5-HT function. It is less certain whether or not major depression is associated with changes in the sensitivity of specific pre- and post-synaptic 5-HT receptor subtypes. However, the blunted hypothermic responses to 5-HT_{1A} receptor agonists suggest a deficit at 5-HT_{1A} raphe autoreceptors in depression. This, however, can not account for a decrease in 5-HT neurotransmission because lowered sensitivity of inhibitory 5-HT_{1A} autoreceptors should produce the opposite effect.

Interestingly, cortisol treatment in healthy subjects also results in blunted hypothermic responses to 5-HT_{1A} receptor agonist challenge [52]. There is evidence from animal experimental studies that corticosterone can modulate both the binding and responsiveness of 5-HT_{1A} receptors in discrete brain regions [53]. It is possible therefore that some of abnormalities seen in response to 5-HT_{1A} receptor challenge in depressed patients are, in fact, secondary to cortisol hypersecretion [54].

Overall it seems likely that major depression is associated with changes in both pre and post-synaptic 5-HT mechanisms. The precise nature of the abnormalities appears to depend on specific patient characteristics and perhaps on the modulatory effect of hormones such as cortisol.

Antidepressant treatment and 5-HT neuroendocrine function

Many antidepressant drugs have prominent acute effects on 5-HT neurotransmission and indeed animal studies have indicated that facilitation of 5-HT neurotransmission, particularly through post-synaptic 5-HT$_{1A}$ receptors may be a common ultimate mechanism of action of many different kinds of antidepressant treatment [55].

The evidence that antidepressants have a common mechanism of action in terms of 5-HT neurotransmission in humans is rather less compelling. For example, the PRL response to intravenous TRP which probably reflects hypothalamic 5-HT$_{1A}$ neurotransmission is enhanced by drugs that block the re-uptake of 5-HT but not by those which act preferentially on NA mechanisms [47]. This is consistent with findings from monoamine precursor depletion strategies which suggest that antidepressants may work through either 5-HT or NA mechanisms to achieve an antidepressant effect [56].

Neuroendocrine tests have also been used to assess neuroadaptive changes in 5-HT neurotransmission during antidepressant treatment. This is because while the ability of antidepressant drugs to increase NA and/or 5-HT function can be detected within hours of drug administration, the therapeutic effect of treatment can take several weeks to become manifest. From this it has been argued that repeated antidepressant treatment results in neuroadaptive changes in 5-HT and other neurotransmitter receptors and it is these adaptive changes that produce the antidepressant effect [55].

In the case of SSRIs, for example, it has been suggested from animal experimental studies that desensitisation of inhibitory cell body 5-HT$_{1A}$ autoreceptors is a necessary pre-requisite before SSRIs can fully increase 5-HT neurotransmission [55]. The effects of SSRI treatment on post-synaptic 5-HT$_{1A}$ receptor function in animal studies are rather contradictory and depend on the model employed. However, neuroendocrine and behavioural studies suggest that SSRIs decrease the sensitivity of post-synaptic 5-HT$_{1A}$ receptors [40].

There is also evidence from behavioural studies that SSRI treatment may decrease the sensitivity of other post-synaptic receptors such as 5-HT$_{2A}$ receptors and 5-HT$_{2C}$ receptors [40]. In addition, there are more limited data to suggest that SSRIs may desensitise 5-HT$_{1B}$ terminal autoreceptors [55]. Like the desensitisation of 5-HT$_{1A}$ autoreceptors, this action would be expected to facilitate pre-synaptic 5-HT release.

From this it can be seen that the net effect of SSRIs on 5-HT neurotransmission is likely to involve a balance between a complex series of neuroadaptative changes in both pre and post-synaptic 5-HT receptors. The desensitisation of 5-HT$_{1A}$ and 5-HT$_{1B}$ receptors would be expected to increase 5-HT neurotransmission while the functional down-regulation of post-synaptic 5-HT$_{1A}$ and 5-HT$_2$ receptors would produce the opposite effect. How far do neuroendocrine challenge tests throw light on this complex situation?

There is good evidence from studies of patients and healthy subjects that repeated treatment with SSRIs decreases the cortisol and GH responses to post-synaptic 5-HT$_{1A}$ receptor stimulation [57]. These studies also indicate that SSRI treatment blunts the hypothermic response to 5-HT$_{1A}$ receptor stimulation [57]. Because this response appears, in part, to be mediated by pre-synaptic 5-HT$_{1A}$ autoreceptors, this suggests that SSRI treatment in humans may desensitise 5-HT$_{1A}$ autoreceptors. Thus these data indicate that SSRI treatment decreases the responsiveness of both pre- and post-synaptic 5-HT$_{1A}$ receptors.

There is less work on the effect of SSRIs on other 5-HT receptor subtypes in humans. We found that the decrease in plasma PRL produced by the 5-HT$_{1B/1D}$ receptor agonist, sumatriptan was not altered by repeated SSRI treatment in healthy subjects [58]. Because sumatriptan-induced decreases in plasma PRL may be mediated by 5-HT$_{1B}$ terminal autoreceptors, this suggests that SSRIs may not desensitise pre-synaptic 5-HT$_{1B}$ receptors in humans.

There have also been some neuroendocrine challenge studies on the effects of SSRI treatment on 5-HT$_{2C}$ receptor sensitivity, using the 5-HT$_{2C}$ receptor agonist, mCPP, as a pharmacological probe. Repeated administration of SSRIs lowers the PRL and hyperthermic responses to mCPP in healthy subjects [59]. This suggests that in humans, as in animals, SSRI treatment lowers the sensitivity of post-synaptic 5-HT$_{2C}$ receptors.

These studies in humans indicate that SSRIs do indeed produce adaptive changes in certain pre- and post-synaptic 5-HT receptors (Tab. 3). While the net effects of the various changes on 5-HT neurotransmission is hard to predict, neuroendocrine studies with TRP suggest that overall SSRIs increase brain 5-HT neurotransmission, particularly at post-synaptic 5-HT$_{1A}$ receptors. This is consistent with the ability of techniques that lower brain

Table 3. Effect of repeated SSRI treatment on 5-HT receptor sensitivity in rat and human

	Rat	Human
5-HT$_{1A}$ pre-synaptic (hypothermia)	↓	↓
5-HT$_{1A}$ post-synaptic (endocrine)	↓	↓
5-HT$_{1B}$ pre-synaptic (electrophysiology/neuroendocrine)	↓	O
5-HT$_{2C}$ post-synaptic (hyperthermia)	↓	↓
5-HT$_{1A}$ overall neurotransmission (electrophysiology/neuroendocrine)	↑	↑

5-HT function, for example, TRP depletion, to produce a rapid reversal of the antidepressant effects of SSRIs in depressed patients [56].

The effects of other antidepressants such as TCAs on neuroadaptive changes in 5-HT receptors have been less studied. However, using buspirone as 5-HT$_{1A}$ receptor probe, we found no effect of the NA re-uptake inhibitor, lofepramine, to alter pre and post-synaptic 5-HT$_{1A}$ receptor function [60]. This is again consistent with data noted above suggesting that antidepressant drugs probably have differing modes of action and may not achieve their therapeutic effects though a final common pathway.

Acknowledgement
The studies of the author described in this review were supported by the Medical Research Council.

References

1. Sedvall G, Farde L, Persson A, Wiesel FA (1986) Imaging of neurotransmitter receptors in the living human brain. *Arch Gen Psychiatry* 43: 995–1005
2. Laruelle M, Abi-Darghan A, Vandyck CH, Rosenblatt W, Zea-Ponce Y, Zoghbi SS, Baldwin RM, Charney DS, Hoffer PB, Kung HF, et al (1995) SPECT imaging of striatal dopamine release after amphetamine challenge. *J Nucl Med* 36: 1182–1190
3. Checkley SA (1992) Neuroendocrinology. In: ES Paykel (ed): *Handbook of affective disorders*. Churchill Livingstone, Edinburgh, 255–266
4. Cowen PJ (1993) Serotonin receptor subtypes in depression: evidence from studies in neuroendocrine regulation. *Clin Neuropharmacol* 16 (Suppl 3): S6–S18
5. Coupland N, Glue P, Nutt DJ (1992) Challenge tests: assessment of the noradrenergic and GABA systems in depression and anxiety disorders. *Mol Aspects Med* 13: 221–247
6. O'Keane V, O'Flynn K, Lucey J, Dinan TG (1992) Pyridostigmine-induced growth hormone responses in healthy and depressed subjects: evidence for cholinergic supersensitivity in depression. *Psychol Med* 22: 55–60
7. Laakmann G, Zygan K, Schoen HW, Weiss A, Wittmann M, Meissner R, Blaschke D (1986) Effects of receptor blockers (methysergide, propranolol, phentolamine, yohimbine and prazosin) on desipramine-induced pituitary hormone stimulation in humans. 1: Growth hormone. *Psychoneuroendocrinology* 11: 447–461
8. Laakmann G, Wittmann M, Schoen HW, Zygan K, Weiss A, Meissner R, Mueller OA, Stalla GK (1986) Effects of receptor blockers (methysergide, propranolol, phentolamine, yohimbine and prazosin) on desipramine-induced pituitary hormone stimulation in humans. 3: Hypothalamo-pituitary-adrenocortical axis. *Psychoneuroendocrinology* 11: 475–489
9. Cowen PJ, Fraser S, Sammons R, Green AR (1983) Atenolol reduces nocturnal plasma melatonin concentration. *Br J Clin Pharmacol* 15: 579–581
10. Cowen PJ, Fraser S, Grahame-Smith DG, Green AR, Stanford C (1983) The effect of chronic antidepressant administration on β-adrenoceptor function of the rat pineal. *Br J Pharmacol* 78: 89–96
11. Checkley SA, Corn TH, Glass IB, Burton SW, Burke CA. (1986) The responsiveness of central alpha$_2$-adrenoceptors in depression. In: JFW Deakin (ed): *The biology of depression*. Gaskell Press, London, 100–120
12. Mitchell P, Parker G, Wilhelm K, Brodaty H, Boyce P, Hickie I (1991) Growth hormone and other hormonal responses to clonidine in melancholic and non-melancholic depressed subjects and controls. *Psychiatry Res* 37: 179–193
13. Katona CL, Healey D, Paykel ES, Theodorou AE, Lawrence KM, Whitehouse A, White B, Horton RW (1993) Growth hormone and physiological responses to clonidine in depression. *Psychol Med* 23: 57–63
14. Siever LJ, Trestman RL, Coccaro EF, Bernstein D, Gabriel SM, Owen K, Moran M, Lawrence T, Rosenthal J, Horvath TB (1992) The growth hormone response to clonidine in acute and remitted depressed male patients. *Neuropsychopharmacology* 6: 165–177

15. Coote M, Wilkins A, Werstiuk ES, Steiner M (1998) Effects of electroconvulsive therapy and desipramine on neuroendocrine responses to the clonidine challenge test. *J Psychiatr Neurosci* 23: 172–178
16. Schittecatte M, Charles G, Machowski R, Wilmotte J (1989) Tricyclic washout and growth hormone response to clonidine. *Br J Psychiatry* 154: 853–863
17. Price LH, Charney DS, Rubin L, Heninger GR (1986) Alpha$_2$-adrenergic receptor function in depression: the cortisol response to yohimbine. *Arch Gen Psychiatry* 43: 849–858
18. Lesch KP, Laux G, Erb A, Pfuller H, Beckmann H (1987) Growth hormone (GH) response to GH releasing hormone in depression: correlation with GH release following clonidine. *Psychiatry Res* 25: 301–310
19. Thomas R, Beer R, Harris B, John R, Scanlon M (1989) GH responses to growth hormone releasing factor in depression. *J Affect Disord* 16: 133–137
20. Mitchell PB, Bearn JA, Corn TH, Checkley SA (1988) Growth hormone response to clonidine after recovery in patients with endogenous depression. *Br J Psychiatry* 152: 34–38
21. Thompson C, Franey C, Arendt J, Checkley, SA (1988) A comparison of melatonin secretion in depressed patients and normal subjects. *Br J Psychiatry* 152: 260–265
22. Sekula LK, Lucke JF, Heist EK, Czambel RK, Rubin RT (1997) Neuroendocrine aspects of primary endogenous depression. XV. Mathematical modelling of nocturnal melatonin secretion in major depressives and normal controls. *Psychiatry Res* 24: 143–153
23. Katona CLE, Theodorou AE, Horton RW (1987) Alpha-2-adrenoceptors in depression. *Psychiatr Dev* 5: 129–150
24. Heal DJ, Prow MR, Buckett WR (1991) Effects of antidepressant drugs and electroconvulsive shock on pre- and post-synaptic alpha$_2$ adrenoceptor function in the brain: rapid down-regulation by sibutramine hydrochloride. *Psychopharmacology* 103: 251–257
25. Corn T, Thompson C, Checkley SA (1984) Effects of desipramine treatment upon central adrenoceptor function in normal subjects. *Br J Psychiatry* 145: 139–145
26. O'Flynn K, O'Keen V, Lucey JV, Dinan TG (1991) Effect of fluoxetine on noradrenergic mediated growth hormone release: a double-blind placebo-controlled study. *Biol Psychiatry* 30: 377–382
27. Langer SZ (1997) Twenty-five years since the discovery of pre-synaptic receptors: present knowledge and future perspectives. *Trends Pharmac Sci* 18: 95–99
28. Thompson C, Mezey G, Corn T, Franey C, English J, Arendt J, Checkley SA (1985) The effects of desipramine upon melatonin and cortisol secretion in depressed and normal subjects. *Br J Psychiatry* 147: 389–393
29. Cowen PJ, Green AR, Grahame-Smith DG, Braddock LE (1985) Plasma melatonin during desmethylimipramine treatment: evidence for changes in noradrenergic transmission. *Br J Clin Pharmacol* 19: 799–805
30. Demisch K, Demisch L, Nickelsen T (1986) Melatonin and cortisol increase after fluvoxamine. *Br J Clin Pharmacol* 22: 620–622
31. Childs PA, Rodin I, Martin NJ, Allen NHP, Plaskett L, Smythe PJ, Thompson C (1995) Effect of fluoxetine on melatonin in patients with seasonal affective disorder and matched controls. *Br J Psychiatry* 166: 196–198
32. Tuomisto J, Mannisto P (1985) Neurotransmitter regulation of anterior pituitary hormones. *Pharmacol Rev* 37: 249–332
33. Meltzer HY, Kolakowska T, Fang VS, Fogg L, Robertson A, Lewine R, Strahilevitz M, Bush D (1984) Growth hormone and prolactin response to apomorphine in schizophrenia and major affective disorders: relation to duration of illness and affective symptoms. *Arch Gen Psychiatry* 41: 512–519
34. Ansseau M, Vonfrenckel L, Cerfontaine R (1988) Blunted response of growth hormone to clonidine and apomorphine in endogenous depression. *Br J Psychiatry* 153: 65–71
35. Pitchot W, Hansenne N, Gonzalez-Moreno A, Ansseau M (1995) Effect of previous antidepressant therapy on the growth hormone response to apomorphine. *Neuropsychobiology* 32: 19–22
36. Matussek N (1988) Catecholamines and mood: neuroendocrine aspects. *Curr Top Neuroendocrinology* 8: 145–182
37. Anderson IM, Cowen PJ (1991) Prolactin response to the dopamine antagonist, metoclopramide, in depression. *Biol Psychiatry* 30: 313–316

38. Ainsworth K, Smith SE, Zetterstrom TSC, Pei Q, Franklin M, Sharp T (1998) The effects of antidepressant drugs on dopamine D_1 and D_2 receptor expression and dopamine release in the nucleus accumbens of the rat. *Psychopharmacology* 140: 470–477
39. Cowen PJ, Braddock L E, Gosden B (1984) The effect of amitriptyline on the growth hormone response to apomorphine. *Psychopharmacology* 83: 378–379
40. Cowen PJ (1998) Pharmacological challenge tests and brain serotonin function in depression and during SSRI treatment. In: M Briley, S Montgomery (eds): *Antidepressant therapy at the dawn of the third millennium*. Martin Dunitz, London, 175–189
41. Lesch KP (1992) 5-HT_{1A} receptor responsivity in anxiety disorders and depression. *Prog Neuropsychopharmacol Biol Psychiatry* 15: 723–733
42. Seletti B, Benkelfat C, Blier P, Annable L, Gilbert F, de Montigny C (1995) Serotonin$_{1A}$ receptor activation by flesinoxan in humans: body temperature and neuroendocrine responses. *Neuropsychopharmacology* 13: 93–104
43. Bill DJ, Knight M, Forster EA, Fletcher A (1991) Direct evidence for an important species difference In the mechanism of 8-OH-DPAT-induced hypothermia. *Br J Pharmacology* 103: 1857–1864
44. Murphy DL, Mueller EA, Hill JL, Tolliver TJ, Jacobsen FM (1989) Comparative anxiogenic neuroendocrine and other physiologic effects of m-chlorophenylpiperazine given intravenously or orally to healthy volunteers. *Psychopharmacology* 98: 275–282
45. Seibyl JP, Krystal JH, Price LH, Woods SW, D'Amico C, Heninger GR, Charney DS (1991) Effects of ritanserin on the behavioural, neuroendocrine and cardiovascular responses to meta-chlorophenylpiperazine in healthy human subjects. *Psychiatry Res* 38: 227–236
46. Whale R, Cowen PJ (1998) Probing the function of 5-$HT_{1B/1D}$ receptors in psychiatric patients. *CNS Spectrums* 3: 40–45
47. Power AC, Cowen PJ (1992) Neuroendocrine challenge tests: assessment of 5-HT function in anxiety and depression. *Mol Aspects Med* 13: 205–220
48. Newman ME, Shapira B, Lerer B (1998) Evaluation of central serotonergic function in affective and related disorders by the fenfluramine challenge test: a critical review. *Int J Neuropsychopharmacol* 1: 49–69
49. Coccaro EF, Siever LJ, Klar HM, Maurer G, Cochrane K, Cooper TB, Mohs RC, Davis KL (1989) Serotonergic studies in patients with affective and personality disorders. *Arch Gen Psychiatry* 46: 587–599
50. Cowen PJ, Power AC, Ware CJ, Anderson IM (1994) 5-HT_{1A} receptor sensitivity in major depression: a neuroendocrine study with buspirone. *Br J Psychiatry* 164: 372–379
51. Meltzer HY, Maes M (1995) Effects of ipsapirone on plasma cortisol and body temperature in major depression. *Biol Psychiatry* 38: 450–457
52. Young AH, Sharpley AL, Campling GM, Hockney RA, Cowen PJ (1994) Effects of hydrocortisone on brain 5-HT function and sleep. *J Affect Disord* 32: 139–146
53. Beck SG, Choi KC, List TJ, Okuhara DY, Birnstiel S (1996) Corticosterone alters 5-HT_{1A} receptor-mediated hyperpolarization in area CA1 hippocampal pyramidal neurons. *Neuropsychopharmacology* 14: 27–33
54. Dinan TG (1994) Glucocorticoids and the genesis of depressive illness: a psychobiological model. *Br J Psychiatry* 164: 365–371
55. Blier P, de Montigny C (1994) Current advances and trends in the treatment of depression. *Trends Pharmac Sci* 15: 220–226
56. Delgado PL, Moreno FA, Potter R, Gelenberg AJ (1998) Norepinephrine and serotonin in antidepressant action: evidence from neurotransmitter depletion studies. In: M Briley, S Montgomery (eds): *Antidepressant therapy at the dawn of the third millennium*. Martin Dunitz, London, 141–161
57. Sargent P, Williamson DJ, Pearson G, Odontiadis J, Cowen PJ (1997) Effect of paroxetine and nefazodone on 5-HT_{1A} receptor sensitivity. *Psychopharmacology* 132: 296–302
58. Wing Y-K, Clifford EM, Sheehan BD, Campling GM, Hockney RA, Cowen PJ (1996) Paroxetine treatment and the prolactin response to sumatriptan. *Psychopharmacology* 124: 377–379
59. Quested DJ, Sargent PA, Cowen PJ (1997) SSRI treatment decreases prolactin and hyperthermic responses to mCPP. *Psychopharmacology* 133: 305–308
60. Herdman JRE, Cowen PJ, Campling GM, Hockney RA, Laver D, Sharpley AL (1993) Effect of lofepramine on 5-HT function and sleep. *J Affect Disord* 29: 63–72

Brain cytokines and the psychopathology of depression

Brian E. Leonard

Pharmacology Department, National University of Ireland, Galway, Ireland

Introduction

Ever since the time of the Roman physician Galen in 200 AD, evidence has been provided to indicate that bacterial and viral infections, autoimmune diseases, cancers as well as predisposition to heart disease frequently increase in patients with major depression [1]. Previously such phenomena were thought to be coincidental partly because there was a lack of a theoretical framework whereby such diverse phenomena could be associated with an abnormality in brain function which is presumed to cause depression. However, during the past 20 years, evidence has accumulated showing that profound changes occur in both cellular and non cellular immunity in patients with major depression and that such changes could be causally linked to the onset of physical ill health [2]. While the mechanisms linking these phenomena are complex, stress undoubtedly plays a key role in triggering many of the changes. It is known that corticotrophin releasing factor (CRF) is elevated in the cerebrospinal fluid of untreated depressed patients which is associated with the hypercortisolaemia that is a common feature of major depression [3]. The rise in cortisol causes a suppression of some aspects of cellular (e.g. natural killer cell, T-lymphocyte replication) immunity [15] which might contribute to the vulnerability of the patient to various infections. However, not all aspects of cellular immunity are suppressed in the depressed patient. Thus the release of proinflammatory cytokines (such as interleukins 1 and 6 (IL-1,6) and tumour necrosis factor alpha (TNF-α) are increased and associated with an elevated macrophage activity. It has been hypothesized that the increases in the proinflammatory cytokines in the brain are major factors causing depression; this forms the basis of the macrophage theory of depression [4]. Thus depression may be perceived as both causing changes in the immune system and also of being caused by immune changes.

Cytokines an the central nervous system

Cytokines are relatively low molecular weight (250–500) polypeptides that exhibit autocrine, paracrine and endocrine properties and which elicit their physiological responses by activating specific receptors. The cytokines are involved in the control of cell growth, inflammatory processes, cell differentiation and the regulation of hormone secretion and the immune system.

Cytokines are produced by monocytes, lymphocytes, mast cells, fibroblasts, endothelial and mesangial cells as well as by astrocytes and microglial cells in the brain. Some 18 different interleukins have so far been identified which, together with other peptides such as the interferons and tumour necrosis factors, comprise a complex network of immuno transmitters [5]. Because of the potential importance of the proinflammatory cytokines in the aetiology of depression, this review will concentrate primarily on the monocyte derived interleukins (IL-1, IL-6) and TNF. These factors are early participants on the immune response and therefore serve as a signal to the brain that an antigen is being processed by the immune system. In addition, IL-1 is known to be a potent activator of the pituitary-adrenal axis further suggesting that specific cytokines play a major role in communicating between the immune system and the brain. Furthermore, cytokines participate in the activation of the peripheral sympathetic system. As the synthesis and release of cytokines occur in response to both exogenous stress and to immune challenges, it is apparent that their synthesis occurs in response to stimuli that threaten the integrity of the organism. Such stimuli presumably include those that initiate depression, environmental stress being a major factor in this regard.

In addition to the evidence that the astrocytes and microglia in the brain form part of the immune system, it is also apparent that the cytokines produced peripherally can communicate with the brain thereby enabling homeostasis to be achieved *via* the release of neuropeptides and hormones. The evidence for a communication between the immune system and the brain, and *vice versa*, is provided by:

(1) an increase in the plasma catecholamine concentration following the injection of a T-dependent antigen [6]. This suggests that the activated immune system signals the brain which results in the increased sympathetic activity.
(2) The EEG response to an immune response [7]. Thus electrodes recording from the pre-optic area of the anterior hypothalamus and dorsal paraventricular nucleus of rats show changes in electrical activity that reflect the immune response to a challenge with sheep erythrocytes.
(3) IL-1 is a potent activation of the hypothalamic-pituitary adrenal (HPA) axis [8, 9], a property that is shared by the other pro-inflammatory

cytokines but not by IL-2 or interferon gamma (IFN-γ). This would appear to be due to a direct effect of the pro-inflammatory cytokines on the release of corticotrophin releasing factor (CRF).
(4) Fever is caused by proinflammatory cytokines altering the thermoregulatory properties of the preoptic anterior hypothalamus [10]. These cytokines function by initiating the release of prostaglandins and possibly CRF.
(5) Proinflammatory cytokines can cause lethargy and sleepines [11].

Thus many of the features that characterize depression are associated with an increase in proinflammatory cytokines in the brain. These include anorexia, disturbed sleep pattern, anhedonia, loss of libido and depressed mood.

The mechanisms whereby the cytokines enter the brain from the periphery are still the subject of debate. One possibility is that the cytokines pass into the brain *via* the circumventricular organs at the base of the fourth ventricle near the opening of the central canal. While the process of passive diffusion is unlikely to account for the full physiological impact of the pro-inflammatory cytokines on the brain, there is evidence of an active transport mechanism that operates from the vasculature to the CSF which could play a role [12]. Retrograde transport *via* the vagus nerve is another possibility. Evidence for this phenomena is provided by the observation that the behavioural effects of intraperitoneally administered IL-1 or TNF-α can be prevented by the severing of the afferent branches of the sub-diaphragmatic portion of the vagus [13, 14]. The precise mechanism of transport is unclear but it could involve receptors on paraganglia located adjacent to the hepatic branch of the afferent vagus.

Cytokines and depression

Evidence implicating a role for the cytokines in the aetiology of depression has been provided by studies of the effects of IL-1, IL-2, TNF-α and interferon-α (INF-α) on psychiatrically normal individuals being treated for a malignancy. In addition, studies of the immune system in depressed patients has provided convincing evidence that cytokines may produce profound changes in the mood state. These two areas will be considered in support of the macrophage theory of depression which implicates an activation of the macrophages, with the consequent release of proinflammatory, cytokines, as the cause of depression [4].

The administration of IL-1, IL-2, TNF-α or INF-α to psychiatrically normal individuals frequently results in depressed mood, anxiety, cognitive impairment, lack of motivation for work and family and loss of libido. These symptoms occur rapidly following administration and usually disappear once the cytokine has been withdrawn [16]. Such behavioural changes appear to be a consequence of the neurotransmitter and endocrine changes induced by the cytokines rather than being induced by the pathological condition (for example, cancer) for which the treatment is administered [16]. In addition to the direct effects of cytokines on behaviour, there is also evidence that patients recovering from various infections frequently exhibit symptoms of depression [17]. Similarly, a high incidence of depression has been reported to occur in patients with multiple sclerosis [18], allergies [19] and rheumatoid arthritis [20]. There is abundant clinical evidence that such cytokines are over-expressed in auto-immune disease [21]. The occurrence of depression is well known to be at least twice as prevalent in woman than men. Women exhibit a greater degree of immune activation than men [22] which might predispose them to all types of depression including post natal depression. The results of these clinical studies suggest that some proinflammatory cytokines can initiate many of the symptoms of depression in otherwise psychiatrically normal individuals.

The initial studies linking depression with an abnormality of the immune system indicated that there was an impairment in neutrophil phagocytosis [23], impaired mitrogen stimulated lymphocyte proliferation [24] and reduced natural killer cell (NKC) activity [25] in untreated depressed patients. These changes in cellular immunity largely return to normal values following effective antidepressant treatment. Stressful life events, such as bereavement, divorce and examination stress in university students, has also been reported to cause qualitatively similar changes in cellular immunity to those occurring in major depression [26]. Such changes appear to be related to the ability of the individual to cope with the stressful event [27]. This raises the question whether the changes in cellular immunity that occur in depression are a consequence of an increased sensitivity of the depressed patient to environmental stressors. While this is a plausible hypothesis, there is evidence that an activation of the immune system can occur in depressed patients that is unrelated to a stressful environmental event [28]. Thus whereas the traditional monoamine hypothesis of depression implicates a disorder in biogenic amine function as the cause of depression, it is now postulated that the changes in neurotransmitter function are secondary to those occurring in the hypothalamic-pituitary-adrenal axis, a change which reflects the activation of both central and peripheral macrophages to produce proinflammatory cytokines.

Recent research into the immune changes occurring in depressed patients has concentrated on the determination of cytokines, soluble cytokine receptors and acute phase proteins in the plasma of patients. Thus Song et al. [29] have reported that the plasma concentration of positive acute phase proteins (α1-acid glycoproteins, haptoglobin, α1-antitrypsin, α1 and α2 macroglobulin) were raised in depressed patients while the negative acute phase proteins (e.g. albumin) were reduced. These changes in acute phase proteins are a reflection of the action of IL-1 and IL-6 on the liver. In addition, Song et al. [29] also showed that the complement proteins (C_3, C_4

and the immunoglobulin IgM) are increased in these patients. Such changes are evidence of immune activation involving the proinflammatory cytokines (acute phase proteins) and B-cells (IgM) that are activated by the cytokines.

Further evidence of immune activation occurring in depressed patients is provided by the studies of Sluzewska et al. [30] who showed that the plasma concentration of IL-1, IL-6, IFN-, and the soluble IL-6 and IL-2 receptors and the IL-1 receptor antagonist are raised in patients with major depression; these changes were correlated with the rise in plasma acute phase proteins. Most of these immune changes were found to return to control values following the effective treatment of the patient with an antidepressant [31].

In addition to these studies of the cytokines in depressed patients, there is also evidence that depressed patients have an increased number of T-helper, T-memory, activated T-cells and B-cells which are responsible for the changes in the plasma cytokines and acute phase proteins already referred to [32]. How these changes are linked to those in central neurotransmitter function is unclear but there is evidence that depressed patients display increased serum antibody titres to serotonin [33] which, if present in the brain, could impair central serotonergic function. In addition, there is evidence that the serum concentration of prostaglandins E_1 and E_2 (PGE_1, PGE_2) are raised in depressed patients [34]; these changes are partly initiated by the stimulation of cycloxygenase activity in the brain by proinflammatory cytokines. The rise in the PGE's in the brain may contribute to the central neurotransmitter deficit by reducing the release of the biogenic amines [35]. This may provide the link between immune activation, changes in brain neurotransmitters and the onset of depression.

To date, relatively few studies have examined the effects of antidepressant treatment of these immune changes. Those studies that have followed the response of the depressed patient through treatment have shown that effective antidepressant treatment reduces the activated immune system [30, 36] thereby suggesting that the immune changes are state, rather than trait, markers of depression.

Stress, the immune system and depression

Hypersecretion of CRF has been shown to play a role in the pathophysiology of depression [37]. This results in an activation of the HPA axis and hypercortisolaemia which is a frequent characteristic of patients with major depression. It would appear that the hypercortisolaemia is associated with a decreased sensitivity of the central glucocorticoid receptors that thereby show a reduced response to the inhibitory feed-back mechanism whereby the plasma cortisol would normally inhibit CRF and ACTH release [38]. An additional factor leading to the hypercortisolaemia in depressed patients is due to the enhanced CRF release caused by the proinflammatory cytokines, particularly IL-1 [39]. This raises a paradox. Glucocorticoids are known to suppress macrophage activity and therefore it would be anticipated that in depression cellular immunity would be dramatically reduced as a consequence of the hypercortisolaemia. The explanation is probably related to the decrease in the sensitivity of the glucocorticoid receptors of the immunocytes, a situation that is analogous to the changes in the glucocorticoid receptors in the brain. Chronic antidepressant treatments normalise the functioning of the HPA axis and the immune system by increasing the sensitivity of the glucocorticoid receptors; one possible mechanism whereby this occurs involves the increased translocation of the glucocorticoid receptor from the cytoplasm to the nucleus thereby increasing glucocorticoid receptor gene transcription [40].

In addition to the interconnection between the proinflammatory cytokines and CRF, there is also evidence that CRF (independently of its effect in the HPA axis) produces behavioural, immmune and neurotransmitter changes in animals following its direct administration into the brain [41]. The behaviour changes resemble those observed following severe stress and depression. Because the effects of CRF in rodents were only observed following subchronic (5 days) intracerebroventricular administration of the peptide, it is difficult to extrapolate such findings to the clinically depressed patient in whom the CRF concentration is probably raised throughout the active phase of the condition. Nevertheless, it does indicate the interrelationship between the activation of the HPA and immune system, the increase in the release of the stress hormone CRF and the subsequent changes in behaviour that follow these defects in neurotransmitter function.

Experimental and clinical studies indicate that stress and depression are associated with an increase in proinflammatory cytokines and an activation of the HPA axis. There is also evidence that activation of the immune system by the administration of a macrophage activator such as lipopolysaccharide, or by the administration of one of the proinflammatory cytokines, can induce depressive symptoms. CRF apears to play a role as a mediator of some of the behavioural changes seen following severe stress, and in the depressed patient. Despite the complexity of the interactions between the cytokines, central neurotransmitters and the glucocorticoids there is growing evidence to support the view that the proinflammatory cytokines play a primary role in the aetiology of depression. This forms the basis of the macrophage theory of depression [14]. However, in spite of the convincing evidence that immune changes occur in depressed patients, there are also studies that have failed to show any change in immune parameters in depression. Such inconsistencies may result from a heterogenous population of patients (e.g. mild and severe depression, bipolar depression) studied. In addition, age and gender of the patients may also influence the results. Clearly there is a need to critically control the clinical status of the patients and to standardize the methods used to assess the endocrine and immune status to fully validate the relationship between the hypersecretion of proinflammatory cytokines and depression.

Prostaglandins, the immune system and depression

Comment has already been made on the increased frequency of depression in patients suffering from autoimmune diseases and rheumatoid arthritis [20], conditions which often respond to treatment with tricyclic antidepressants. In addition to the increase in the tissue concentrations of proinflammatory cytokines, prostaglandins of the E series (PGE's) also play a crucial part in such conditions. Furthermore, in major depression, and in the absence of an inflammatory disease process, there is evidence of a substantial increase in the plasma PGE's concentrations [34] and in the synthesis and release of PGE2 from whole blood samples in depressed patients [42]. Thus the possibility arises that the PGE's may provide a link between the inflammatory process that occur in depression and autoimmune disease, changes in the immune system and the symptom of the condition. While the evidence in favour of prostaglandin hypothesis of depression is indirect, it does raise the possibility that a new type of antidepressant could be developed that acts as an inhibitor of prostaglandin synthesis in the brain. So far, the poor lipophilicity of the available non-steroidal antiinflammatory drugs (and their lack of specificity for the cycloxygenases in the brain) has prevented the validation of this hypothesis.

Antidepressants as prostaglandin synthetase inhibitors

Over 20 years ago, it was shown that the tricyclic antidepressant clomipramine inhibited the pressor responses to potassium ions and vasopressin in the rat mesenteric vascular bed, an effect which was attributed to an inhibition of the synthesis of PGE [43]; the NSAID indomethacin was shown to have a similar effect to clomipramine. Furthermore, the concentration of clomipramine which produced this antiprostaglandin effect was within the therapeutic range. This finding led Horrobin [44] to propose that a range of psychotropic drugs with antidepressant activity (for example, the tricyclic antidepressants, monoamine oxidase inhibitors and lithium) act as central cycloxygenase inhibitors. Conversely drugs such as reserpine and alpha-methyl dopa, which can precipitate depression in some hypertensive patients, activate cyclooxygenase; oral contraceptives have a qualitatively similar effect [45]. Such observations provide support for the hypothesis that antidepressants not only modulate central biogenic amine function but also inhibit PGE synthesis both centrally and peripherally. This could account for the apparent antiinflammatory action of the tricyclic antidepressant.

Prostaglandins as modulators of the immune response

Preliminary studies showed that neutrophil phagocytosis is reduced in a state dependent manner in patients with major depression [23]. Such changes would appear to be unique to patients suffering from depression and panic attack as they were not apparent in those with generalized anxiety disorder, schizophrenia, Alzheimer's disease or alcoholism. It was shown that the disorder in neutrophil phagocytosis was associated with a serum factor [23] and subsequent studies indicated that PGE2 was probably responsible [46]. Further evidence for the direct involvement of PGE's in the suppression of neutrophil phagocytosis was provided by the observation that aspirin, administered in therapeutic doses to healthy subjects, enhanced neutrophil phagocytosis. Further confirmation of the role of PGE's in immune suppression is provided by the experimental study in which the lymphoproliferative response of spleen cells to the mitogen concanvalin A is increased by the chronic pretreatment of mice with indomethacin [47].

In addition to changes in neutrophil phagocytosis, it is also evident that the proliferation of T lymphocytes to a mitogen challenge is reduced in patients with major depression [28]. There is *in vitro* data to suggest that PGE2 causes the reduction in T-lymphocyte activity indirectly by inhibiting the synthesis of IL-2 and down regulates the expression of the transferrin receptor on the T lymphocyte surface [48]. Furthermore, the reduction in natural killer cell (NKC) activity, which is a characteristic feature of patients with major depression [25], has also been shown to occur following the *in vitro* incubation of NKC's with PGE [49]. The precise mechanism whereby the PGE's suppress cellular immunity is uncertain but over two decades ago, Mtabaji and co-workers [43] had shown that PGE2 impaired the influx of calcium ions into rat mesenteric vascular tissue, an effect which could be antagonized by the tricyclic antidepressant clomipramine. Subsequently, *in vivo* studies demonstrated that pre-treatment of rats with the cycloxygenase inhibitor indomethacin prevented the PGE2 related depression of Con-A induced T-cell calcium mobilization. This suggests that the PGE2 suppression of T-cell activation is due to an attenuation of calcium signalling. In addition to the effects of NSAID's and clomipramine as the suppression of cellular immunity, anti-PGE antibodies have also been shown to prevent PGE induced immunosuppression [50].

The complement system is one of the major effectors of the non specific humoral immune systems. Song et al. [29] have shown that complement C3 and C4 is raised in patients with major depression. It would appear that these changes are attributable to the increase in the plasma concentration of PGE2 as there is evidence to show that PGE2 increases the C3 concentration. Thus the PGE's appear to be responsible not only for the suppression of many aspects of cellular immunity (NKC's, T-lymphocytes, neutrophils) but also for an increase in some aspects of non specific immunity by increasing the concentration of some complement factors.

An increase in aspects of cellular immunity in the depressed patient is also indicated by the increased release of proinflammatory cytokines already referred to above. *In vitro* studies on the functional and secretory activites of human mononuclear phagocytes have demonstrated that the inhibition of antibody dependent cell cytotoxicity and the secretion of reactive oxygen intermediates by peritoneal macrophages correlated with an increased release of PGE2 [51]. These studies suggest that the inflammatory macrophages whose activities are increased in depression not only contribute to the central changes by the release of proinflammatory cytokines but also increase the local synthesis and release of PGE2 that enhances the inflammatory process.

Conclusion

Despite the substantial evidence that implicates a disorder of the immune system with depression, controversy exists regarding the causal relationship between the increase in proinflammatory cytokines, PGE's and the biological basis of the disease. Are the changes in the immune system a reflection of a stress induced pathological state or are they responsible for the changes in central neurotransmitter function that, following a stress related activation of the immune system, causes the behavioural changes? Only more extensive research in the field of psychoneuroimmunology will help to unravel the complex interrelationship between the immune-endocrine-neurotransmitter systems and the symptoms of depression. A summary of the present situation is shown in Figure 1.

However, consideration of these factors has been important in that it may stimulate the development of new concepts regarding the biology of depression and how antidepressants bring about these therapeutic effects. Perhaps the time has come to move from the conceptually restrictive monoamine theory of depression to novel approaches that take into account the more general changes in the periphery as well as the brain. Could centrally acting PGE inhibitors and/or proinflammatory cytokine antagonists become the antidepressants of the new millennium? Only time, and more research, will tell!

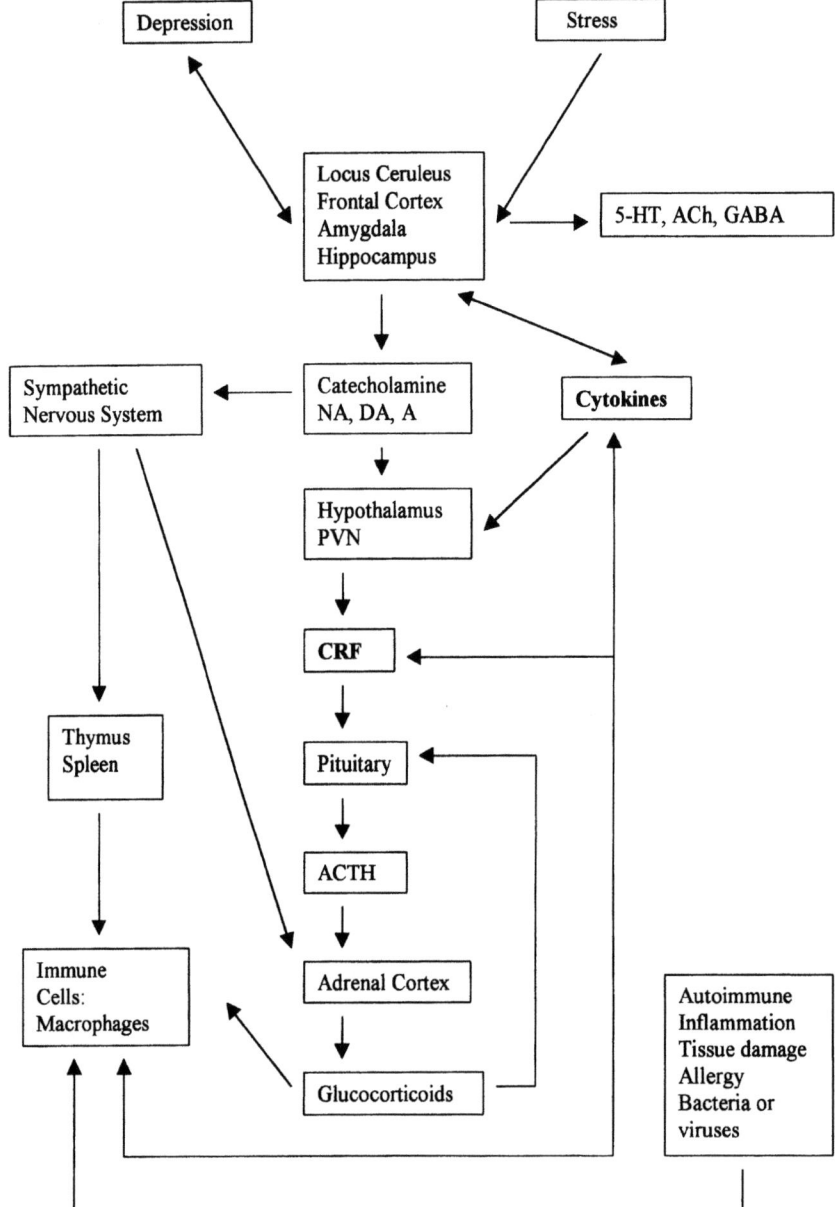

Figure 1. The relationship between cytokines, CRF, neurotransmitters, glucocorticoids and immunity in depression and stress.

References

1 Shaprio PA, Lidagoster L, Glassman AH (1997) Depression and heart disease. *Psychiat Ann* 27: 347–352
2 Dorian B, Garfinkel PE (1987) Stress, immunity and illness: A review. *Psychol Med* 17: 393–407
3 Nemeroff CB, Widerlov E, Bissette G (1984) Elevated concentrations of CSF corticotropin-releasing factor-like immunoreactivity in depressed patients. *Science* 226: 1342–1343
4 Smith RS (1991) The macrophage theory of depression. *Med Hypothesis* 35: 298–306
5 Mire-Sluis A, Thorpe R (eds) (1998) Cytokines. Academic Press, New York
6a Besedovksy HO, Sarkin E (1977a) Hormonal control of immune process. *Endocrinology* 2: 504–513
6b Besedovksy HO, Sarkin E (1977b) Network of immune-neuroendocrine interactions. *Clin Exp Immunol* 27: 1–12
7 Saphier D, Abramsky O, Mor G, Ovadia H (1987) Multiunit electrical activity in conscious rats during an immune response. *Brain Behav Immun* 1: 40–51
8 Besedovksy HO, Del Rey A, Sarkin E, Dinarello CA (1986) Immunoregulatory feedback between interleukin-1 and glucocorticoid hormone. *Science* 233: 652–654
9 Besedovksy HO, Del Rey A, Klusman I (1991) Cytokines as modulators of the hypothalamus pituitary-adrenal axis. *J Steroid Biochem Mol Biol* 40: 613–618
10 Kluger MJ, Kozak W, Leon LR, Soszynski D, Conn CA (1985) Cytokines and fever. *Recent Prog Neuroimmunomod* 2: 216–223
11 Darko DF, Mitler MM, Henriksen SJ (1995) Lentivirus infection, immune response peptides and sleep. *Adv Neuroimmunol* 5: 57–77
12 Banks WA, Kastin AJ, Broodwell RD (1995) Passage of cytokines acros the blood-brain barrier. *Recent Prog. Neuroimmunomod* 2: 241–248
13 Goehler LE, Busch CR, Tartaglia N, Reltan J, Sisk O (1995) Blockade of cytokine induced conditioned taste aversion by subdiaphragmatic vagotomy: further evidence for vogal mediation of immune-brain communication. *Neurosci Lett* 185: 163–166
14 Bluthe RM, Dantzer R, Kelley KW (1997) Central mediation of the effects of interleukin 1 on social explanation and body weight in mice. *Psychoneuroendocrinol* 22: 1–11
15 Song C, Leonard BE (1995) The effect of olfactory bulbectomy in the rat, alone or in combination with antidepressants and endogenous factors on immune function. *Hum Psychopharmac* 10: 7–18
16 Meyers CA, Valentine AD (1995) Neurological and psychiatric adverse effects of immunological therapy. *CNS Drugs* 3: 56–68
17 Yirmiya R (1996) Endotoxin produces a depressive-like episode in rats. *Brain Res* 711: 163–174
18 Minden SL, Schiffer RB (1990) Affective disorders in multiple sclerosis in review and recommendations for clinical research. *Arch Neurol* 47: 98–104
19 Marshall PS (1993) Allergy and depressions: a neurochemical threshold model of the relation between the illnesses. *Psychol Bull* 113: 23–43
20 Katz PP, Yelin EH (1993) Prevalence and correlates of depressive symptoms among persons with rheumatoid arthritis. *J Rheumatol* 20: 790–796
21 Schrott LM, Crnic LS (1996) The role of performance factors in the active avoidance-conditioning deficit in autoimmune mice. *Behav Neurosci* 110: 486–497
22 Grossman CJ (1985) Interaction between the gonadal steroids and the immune system. *Science* 227: 257–261
23 O'Neill B, Leonard BE (1991) Abnormal zymosan-induced neutrophil chemiluminescence as a marker of depression. *J Affect Dis* 9: 265–272
24 Kronfol Z, House JD (1989) Lymphocyte mitogenesis, immunoglobulin and complement levels in depressed patients and normal controls. *Acta Psychiat Scand* 80: 142–147
25 Irwin M, Lacher U, Caldwell C (1992) Depression and reduced natural killer cell cytotoxicity: a longitudinal study of depressed patients and control subject. *Psychol Med* 22: 1045–1050
26 Irwin M (1995) Psychoneuroimmunology of depression. In: KE Bloom, DJ Kupfer (eds): *Psychopharmacology: the fourth generation of progress.* Rowan Press, New York, 983–998

27 Irwin M, Daniels M, Smith TH, Bloom E, Weiner H (1997) Impaired natural killer cell activity during bereavement. *Brain Behav Immun* 1: 98–104
28 Maes M, Smith R, Scharpe S (1995) The monocyte-T lymphocyte hypothesis of major depression. *Psychoneuroendocrinol* 20: 111–116
29 Song C, Dinan T, Leonard BE (1994) Changes in immunoglobulin; complement and acute phase protein levels in depressed patients and normal controls. *J Affect Dis* 30: 283–288
30 Sluzewska A, Rybakowski J, Bosmans E (1996) Indicators of immune activation in major depression. *Psychiat Res* 64: 161–167
31 Sluzewska A, Rybakowski JK, Laciak M (1995) Interleukin-6 serum levels in depressed patients before and after treatment with fluoxetine. *Ann NY Acad Sci* 762: 474–477
32 Maes M, Smith R, Scharpe S (1995) The monocyte T-lymphocyte hypothesis of major depression. *Psychoendocrinol* 20: 111–116
33 Schott K, Batra A, Klein R, Bartels M, Koch W, Berg P (1992) Increased serum antibody titres to serotonin in depressed patients. *Eur Psychiatr* 7: 209–212
34 Calabrese J, Skwereer RG, Barna B (1986) Depressed, Immunocompetence and prostaglandins of the E series. *Psychiat Res* 17: 44–47
35 Hedquist P (1976) Effects of prostaglandins on autonomic neurotransmission. In: SMM Karim (ed) *Prostaglandins: physiological, pharmacological and pathological aspects*. MTP Press, Lancaster U.K. 37–41
36 McAdams A, Leonard BE (1993) Neutrophil and monocyte phagocytosis in depressed patients. *Prog Neuropsychopharmac Biol Psychiat* 17: 971–984
37 Owens MJ, Nemeroff CB (1991) Physiology and pharmacology of corticotrophin releasing factor. *Pharmac Rev* 4: 425–473
38 Dinan TG (1994) Glucocorticoids and the genesis of depressive illness: a psychobiological model. *Br J Psychiat* 164: 365–371
39 Dunn A (1992) The role of interleukin-1 and tumor necrosis factor alpha in the neurochemical and neuroendocrine responses to endotoxin. *Brain Res Bull* 29: 807–872
40 Pariante CM, Pearce BD, Pisell TL, Owens MJ, Miller AH (1997) Steroid-independent translocation of the glucocorticoid receptor by the antidepressant, desipramine. *Eur Neuropsychopharma* 7: 184
41 Song C, Earley B, Leonard BE (1995) Behavioural, neurochemical and immunological response to CRF administration. *Ann NY Acad Sci* 771: 55–72
42 Song C, Lin A, Banaccorso S, Heide C, Verkerk R et al (1998) The inflammatory response system and the availability of plasma tryptophan in patients with primary sleep disorders and major depression. *J Affect Dis* 49: 211–219
43 Mtabaji JP, Manku MS, Horrobin DF (1977) Actions of the tricyclic antidepressant clomipramine on responses to pressor agents: interactions with prostaglandin E2. *Prostaglandins 1* 4: 125–132
44 Horrobin DF (1977) The role of prostaglandins and prolactin in depression, mania and schizophrenia. *Postgrad Med J* 53 (Suppl 4): 160–165
45 Castracane VD, Jordan VC (1975) The effect of oestrogen and progesterone on uterine prostaglandin synthesis in the ovariestomized rat. *Biol Reproduction* 13: 587–595
46 McAdams C, Leonard BE (1992) Effect of prostaglandin E2 and thromboxane A1 on monocyte and neutrophil phagocytosis *in vitro*. *Med Sci Res* 20: 673–674
47 Malave I, Araujo Z (1986) Differences in the effects of indomethacin and pre incubation on the lymphoproliferative response to concanavalin A of spleen cells from low responder C57 BL/6 and high responder BALB/C mice. *Int J Immunopharmacol* 8: 137–146
48 Choaib S, Welte K, Metelsmann R, Diport B (1985) Prostaglandin E2 acts at two distinct pathways of T lymphocyte excitation: inhibition of interleukin 2 production and down regulation of transferrin receptor expression. *J Immunol* 135: 1172–1179
49 Merrill JE, Myers LLW, Ellison GW (1983) Regulation of natural killer cell cytotoxicity by prostaglandin E in the peripheral blood and cerebrospinal fluid of patients with multiple sclerosis and other neurological diseases. Part 2. Effect to exogenous PGE1, an spontaneous and interferon induced natural killer. *J Neuroimmunol* 4: 239–251
50 Freeman TR, Shelby J (1988) Effect of anti-PGE antibody on cell-mediated immune response in thermally induced mice. *J Trauma* 28: 190–194
51 Gudewicz PW, Frewin MB (1991) Surface contact modulation of inflammatory macrophage antibody dependent cytotoxicity and prostanoid release. *J Cell Physiol* 149: 195–201

The future of antidepressants

Antidepressants
ed. by B.E. Leonard
© 2001 Birkhäuser Verlag Basel/Switzerland

Chemistry and pharmacology of novel antidepressants

John S. Andrews[1] and Roger M. Pinder[2]

[1] *Janssen Research Foundation, Turnhoutseweg 30, 2340 Beerse, Belgium*
[2] *Organon Inc. 375 Mount Pleasant Avenue, West Orange, New Jersey 07059, USA*

Introduction

Depression is a common, recurrent and debilitating condition which carries substantial personal, familial and societal costs. It causes significant suffering, disability and social dysfunction frequently leading to disruption of normal life and work. Depressed patients struggle to recover, often over several months and sometimes even years, and many follow a chronic and remorseless course. The high morbidity is accompanied by a substantial mortality, such that as many as 15% of depressed patients take their own lives. Many more attempt suicide at some time during their lives. The economic impact of depression on society is considerable and is comparable to that of other major disorders such as coronary heart disease. However, depression is often not properly recognised, it exacts costs over a longer period of time and places a particularly heavy burden on employers because of lost productivity. Annual costs have been estimated for the USA at about US $ 44 billion [1, 2].

Although almost half of all patients who contact primary health care services are believed to have current depression, the disease is underdiagnosed and undertreated [3, 4]. Two-thirds of those who seek help are not prescribed treatment, and when drug therapy is prescribed only 25% of these subjects receive antidepressant drugs. Swiss data do suggest that 9% of the male and 25% of the female population has been treated for depression by the age of 30, but less than half of those diagnosed with depression ever receive treatment. Even when appropriate treatment is prescribed, sub-therapeutic dosage, particularly of the older antidepressants, is frequent. The benefits of effectively treating depression far outweigh the considerable risks of leaving depressed patients inadequately, inappropriately or not treated [5].

Current antidepressants

There is no cure for depression. However, symptomatic treatments have existed for more than four decades and they have unequivocally altered the short-term outcome of depressive illness and reduced considerably the risk of morbidity. They include antidepressants for the treatment of depressed patients as well as lithium for those with bipolar disorders who experience mood swings between depression and mania [6, 7]. Older antidepressants of the first generation are still widely prescribed although they have deleterious anticholinergic and cardiovascular side-effects that can lead to poor patient compliance and serious toxicity in overdosage. Examples include the tricyclic antidepressants (TCAs) imipramine and amitriptyline which were the prototypes for a whole family of similar drugs. Early monoamine oxidase inhibitors (MAOIs) like iproniazid and tranylcypromine were plagued by serious and sometimes life-threatening hypertensive crises precipitated by interaction with tyramine-containing foodstuffs. Newer, second generation antidepressants such as mianserin, trazodone and the selective serotonin reuptake inhibitors (SSRIs) lack these drawbacks but they do have their own particular pattern of side-effects. Furthermore, like the older antidepressants, they are effective in only about 70% of depressed patients, less than half of whom experience a full response, and take effect only after 2–4 weeks (Tab. 1). Second generation drugs, particularly the SSRIs, are often perceived as being less effective than TCAs especially in severe depression [8]. The mood stabiliser lithium is also no panacea in the treatment of bipolar disorders, being associated with a variety of adverse reactions which often lead to poor compliance and a failure rate of about 33%. Even among responders, few can expect to achieve complete prevention of episodes [9]. Current alternatives to lithium like carbamazepine and valproate are no more effective and carry an extra burden of toxicity [10]. Antidepressants and mood stabilisers need to be given for long periods because of the recurrent and chronic nature of affective disorders.

Table 1. Characteristics of three generations of antidepressants

Generation	Time of first introduction	Examples	Efficacy in > 70% of patients	Onset of action (< 2 – 3 weeks)	Anticholinergic effects	Cardiovascular effects	Toxic in overdose
First[a]	1958	imipramine iproniazid	–	–	+	+	+
Second[b]	1975	mianserin trazodone SSRIs RIMAs NaSSAs	–	–	–	–	–
Third	> 2000	none	+	+	–	–	–

[a] tricyclic antidepressants (inhibitors of monoamine reuptake) and monoamine inhibitors.
[b] drugs of atypical structure and atypical pharmacology.
– absent.
+ present.

Individual depressed patients vary widely in their response to different drugs, and it is necessary to have at hand a range of medications offering multiple mechanisms of action. There are therefore many antidepressants available and good reasons to develop more [6, 7]. The need for alternatives to current mood stabilisers and existing antidepressants is well recognised. Third-generation antidepressants will be able to rapidly alleviate depression, to substantially alter the course of mood disorders and to largely maintain the majority of patients in remission, while retaining the gains already made in terms of reduced side-effects and greater safety in overdose. A less toxic and more effective lithium mimetic is also required. Research targets for improved drugs have been identified [7, 11], and new pharmacological strategies towards developing novel antidepressants have been described in detail [12]. The prospects for true third-generation drugs are promising, nevertheless, improvements in the latest members of the second generation, and improved augmentation strategies, will play a role in producing slightly faster and broader action treatments than with the first and early second generation drugs.

Current antidepressant drugs of all types act rapidly to raise synaptic levels of catecholamines, particularly of noradrenaline (NA) and serotonin (5-HT). The original and still prevailing concept of the biochemical causes of mood disorders is the biogenic amine hypothesis, which postulates that depression is related to a functional deficiency of NA and/or 5-HT at relevant brain receptors. Despite the general feeling that longer-term adaptive changes in receptor sensitivity may better explain the delayed therapeutic action of antidepressants, it is clear that most drugs do enhance central neurotransmission especially of 5-HT upon chronic administration [13]. Furthermore, the hypothesis has remained the basis for developing augmentation strategies as well as the target for new drug design.

Augmentation strategies are now commonplace when patients fail to respond to one or other antidepressant or where a quick onset of action is desired [14, 15]. Augmentation with triiodothyronine (T3) or lithium is well established, while SSRI-resistance or more particularly delayed onset is now being addressed by addition of the 5-HT1A autoreceptor antagonist pindolol [16]. Combinations of different antidepressant drugs have also been used, but neither MAOIs nor SSRIs have been favoured for augmenting TCAs because of the potential problems of cardiotoxicity and pharmacokinetic interactions respectively. Combinations of drugs acting predominantly *via* noradrenergic mechanisms with those working through 5-HT systems have seemingly led to both faster onset and greater efficacy; fluoxetine with desipramine [17], with mianserin [18, 19], and with yohimbine [20] are examples. It appears that selectivity of action is not the optimal profile for an antidepressant, a view amplified by the many studies suggesting that the classical drugs of multiple action, the TCAs, are sometimes superior to those of selective action like the SSRIs [8], and by the efficacy of electroconvulsive therapy (ECT) in drug refractory depression.

Table 2. Targets for third generation antidepressants

Receptor heterogeneity
G-proteins
Second messengers
Protein phosphorylation
Gene transcription factors
Hormones of the hypothalamic-pituitary-adrenal (HPA) axis
Genetic basis of mood disorders

The twin threads of augmentation and multiple action have come together in the designer's approach to dual action drugs [21, 22]. Although the selective noradrenaline reuptake inhibitor reboxetine represents an exception to the trend [23], recently introduced antidepressants have been dominated by drugs of dual action with effects upon both noradrenergic and serotonergic neurotransmission represented by the serotonin noradrenaline reuptake inhibitors (SNRIs) venlafaxine and milnacipran and the noradrenaline and serotonin specific antidepressant (NASSA) mirtazapine. Two antidepressant mechanisms may really be better than one [24–26]. Thus, all three dual action agents have been shown to be superior in efficacy to fluoxetine in the more severe forms of depression [27–29], while venlafaxine and mirtazapine appear to be faster in onset of action.

Such drugs can be designed because the structural features that determine monoamine reuptake or receptor affinity are reasonably well established [21]. They represent a further extension of the second generation drugs, an extension made possible by refinement of the pharmacological concepts underpinning antidepressant design. The essence of this development has been the realisation that two actions upon neurotransmitters may be better than one for efficacy and that receptor-specific actions are of crucial importance to improved tolerability. Current knowledge on structure-activity relationships will, however, become increasingly irrelevant for many of the new targets (Tab. 2), which are largely intracellular and where membrane transport will play a major role. Correlation of the pharmacology of novel agents with their clinical effects will require and generate a new body of knowledge. True third generation drugs will not be available until well into this century, but it is the purpose of this review to address some of the chemical and pharmacological issues surrounding those novel antidepressants that are in development.

Serotonergic directed antidepressants

Single action drugs: selective serotonin reuptake inhibitors (SSRIs)

Following the success of fluoxetine, numerous SSRIs now compete for the same market. However, it is questionable as to how much benefit additional

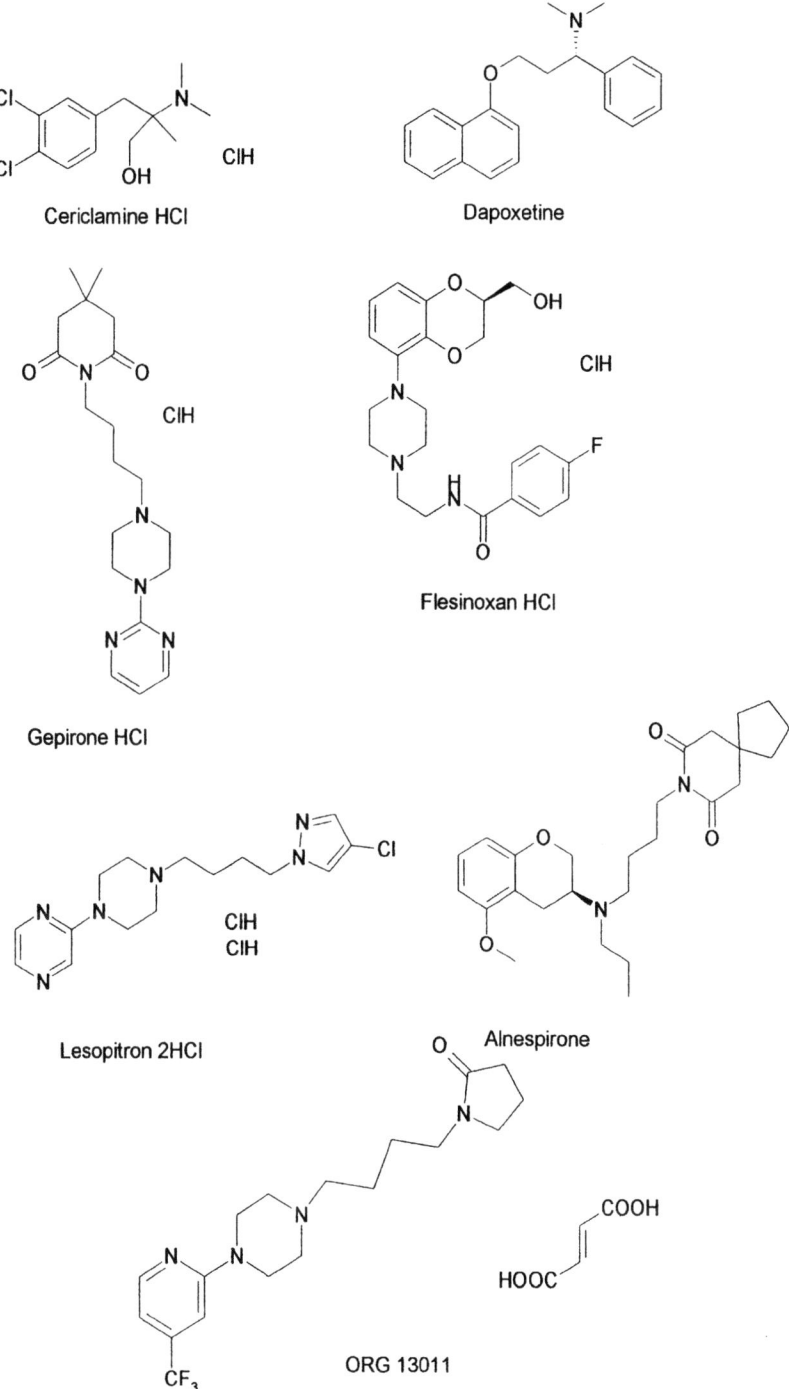

Figure 1A. New antidepressants affecting 5-HT reuptake or 5-HT$_{1A}$ receptors.

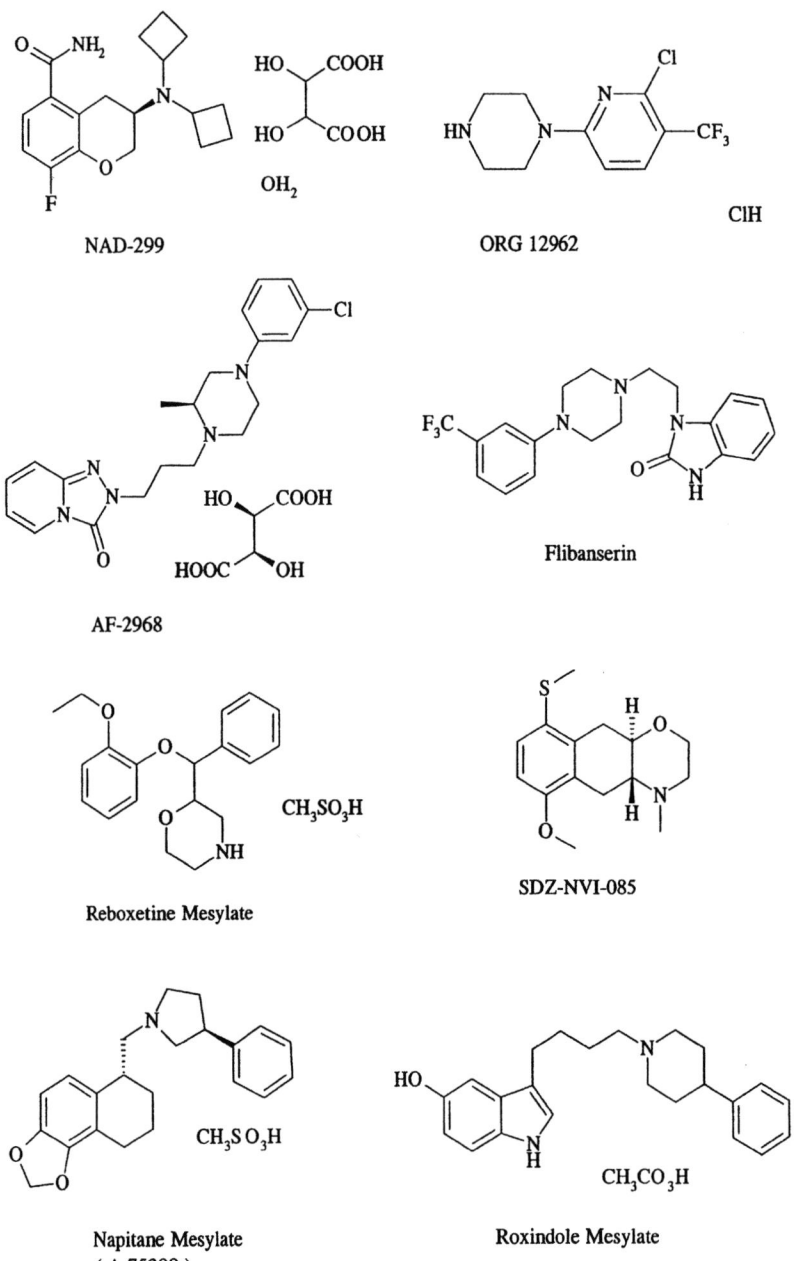

Figure 1B. New antidepressants affecting 5-HT, NA and dopamine mechanisms.

SSRIs can bring to the clinic. Recent SSRIs show greater selectivity than fluoxetine, however, they do not appear anymore effective, quicker in onset or free of troublesome side-effects. It would be reasonable to predict that compounds such as cericlamine and dapoxetine (Fig. 1) will be amongst the last of the SSRIs developed.

Although selective for the serotonin transporter, the SSRI's lack pharmacological specificity. The rapid rise in the use of SSRI's brought an equally rapid appreciation of the pharmacology of serotonin, and the multitude of serotonergic-related side-effects associated with general serotonin activation. Appreciation of specific serotonin-related side-effects and concerns about efficacy in severely depressed patients has led to the development of a more focussed serotonergic approach to depression.

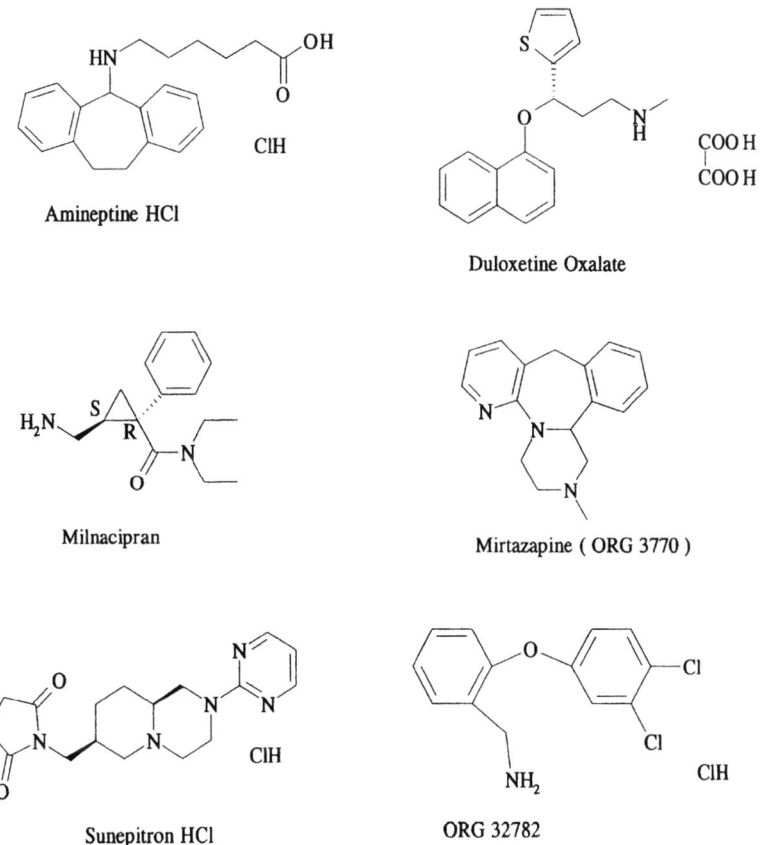

Figure 1C. New antidepressants of dual (and triple) mechanism of action.

Single action drugs: 5HT1A agonists

Several lines of evidence implicated 5HT1A receptors in the therapeutic aspects of SSRI treatment. In addition, research has suggested a mechanism for the delay in action of the SSRIs. Initial serotonin release would feedback at presynaptic 5HT1A autoreceptors and inhibit further release; only after adaptive desensitisation had occurred could the beneficial effects of serotonin release at postsynaptic receptors be felt [19, 24]. Thus, speed of onset could theoretically be improved by bypassing the need for adaptation and directly stimulating postsynaptic 5HT1A receptors. Despite the matureness of the approach, few of the selective compounds synthesised proved successful in clinical trials; the problems encountered were often with pharmacokinetics and side-effects such as dizziness and headache as well as with efficacy.

Figure 1 shows the structures of some of the most advanced compounds in the clinic. Interestingly, initial work with 5HT1A agonists concentrated on anxiety, in part due to the introduction of buspirone (a partial 5HT1A agonist and dopamine antagonist) as an alternative to benzodiazapines. Several compounds with similar structures, but more selective serotonergic properties, followed rapidly. Ipsapirone and gepirone, like buspirone are partial 5HT1A agonists, but lack activity at the dopaminergic receptors; both also share a common metabolite with buspirone (1-(2-pyrimidyinyl)-piperazine) which has alpha2 antagonistic properties [30]. Initial studies indicated efficacy in several animal models of anxiety, but both latter compounds were eventually also tested for antidepressant action. Only gepirone successfully completed Phase III trials for depression and will be submitted for registration shortly [31, 32].

Current reports suggest clinical trials with 5HT1A agonists have refocussed on anxiety. Recent clinical trials for flesinoxan, lesopitron and alnespirone have tended in this direction, although Org 13011 a partial 5HT1A agonist is currently in Phase II for depression [33–35] (Fig. 1). The reasons for this shift are complex: side-effects may have been greater than originally anticipated, however, full agonist therapies have also suffered from the fear of tolerance. In general, continual stimulation of post synaptic receptors tends to a decrease in sensitivity, downregulation of receptors and potentially tolerance. To date there are no reports for gradual dose elevation occurring with the 5HT1A agonists, but the issue remains of general concern to all direct agonist treatments. This concern may favour the ultimate progression of partial agonists such as gepirone which do not elicit such a strong response.

Single action drugs: 5HT1A antagonists

Theoretically a presynaptic inhibitor of 5HT1A receptors should have a similar effect to a 5HT1A agonist. 5HT1A antagonists have proven difficult to optimise and few are close to clinical evaluation. Nevertheless, there are indications that NAD 299 will be evaluated for anxiety and depression [36]. However, the most beneficial role for antagonists may be as augmentation therapies, to remove the initial inhibitory effects of SSRI treatment on presynaptic heteroreceptors and improve or accelerate antidepressant effects. Pindolol was first used to demonstrate the effect [16]; more selective compounds should be of greater benefit.

Single action drugs: 5HT2C agonists

Although much of the medicinal chemistry of agonists has focussed on constructing new 5HT1A agonists, other 5HT receptors may be as important in the therapeutic response. Many of the characteristic effects of fluoxetine are similar in some respects to those of a 5HT2C agonist. For example in animals both fluoxetine and 5HT2C agonists are anorexic, and share similar behavioural effects including discriminative stimulus properties [37].

A useful role for 5HT2C agonists in affective disorders has been much disputed. 5HT2C agonists, such as mCPP, are reported to be anxiogenic in rats. Recent research has questioned this linkage and several studies indicate that 5HT2C agonists may possess anxiolytic and antipanic properties in animals [38, 39]. Moreover, mCPP is not necessarily anxiogenic in man and one open trial has suggested antidepressant efficacy [40]; in fact mCPP is the major metabolite of the atypical antidepressant trazodone. Clinical results are eagerly awaited to settle the dispute and several chemically distinct 5HT2C agonists have been synthesised. The most advanced is currently entering Phase II for the treatment of depression (Org 12962), (Fig. 1) [39]. Meanwhile, 5HT2C antagonists continue to be investigated for their anxiolytic potential [41].

Mixed serotonergic profiles

An alternative approach to a single action drug has been to develop compounds with mixed profiles at 5HT receptors and transporters. Nefazodone was the first SSRI modified to include additional 5HT2 antagonist properties, designed to avoid the potential undesirable aspects of 5HT2 stimulation and leading to a more focussed stimulation of relevant serotonin receptors [42]; nefazadone has also benefited from concomitant pindolol treatment [43].

Another approach has been to develop mixed agonist/antagonists, for example, 5HT1A agonists which are also 5HT2A (flibanserin) [44, 45] (Fig. 1) or 5HT2A and 5HT2C antagonists such as AF 2968 (currently completing Phase I), thereby enhancing 5HT1A activation and reducing unwanted side-effects related to stimulation at other subtypes. Several claims have been issued for a faster onset of action based on effects in animal models, but relevant clinical data is lacking.

Noradrenergic directed antidepressants

Single action drugs: noradrenergic reuptake inhibitors (NRIs)

Research into serotonin transporters as a method of enhancing monoamine levels in the brain has dominated the last two decades. However, modulation of the noradrenergic system appears equally beneficial in depression, regarding the tricyclic NRI desipramine or the alpha-2 blocker mianserin. Reboxetine is one of the few new NRIs to appear since desipramine [46] and has a more selective pharmacological profile, lacking, for example, confounding actions at histamine or muscarinic receptors. Nor does reboxetine suffer from the same side-effect profile as the SSRIs or TCAs; but neither is it side-effect free. Indeed this class suffers from a similar problem to the SSRIs, a lack of pharmacological specificity leading to indiscriminate activity at all noradrenergic receptors. In addition, no studies have yet shown a faster onset of action for selective noradrenergic drugs.

Single action drugs: alpha-2 ligands

A more selective modulation of the noradrenergic system can be achieved by the use of specific alpha-2 antagonists. Some of the earliest atypical antidepressants, such as mianserin, were selective alpha-2 blockers [18, 19]. The difficulty of separating effects on the peripheral from the central nervous system has caused some problems in the development of novel alpha-2 antagonists, however concerns about efficacy with respect to TCAs have also surfaced. The development of new alpha-2 antagonists for depression has effectively stopped, and although a few compounds were reported to be in development such as MK 912 [47] (Fig. 1), most alpha-2 antagonists are aimed at vascular conditions.

The aim of the alpha-2 antagonists was to enhance noradrenaline release by blocking the adrenergic auto and heteroreceptors situated on the synaptic terminal. Alternative strategies could also include selective agonists such as SDZ NVI085, an alpha-1 agonist from Novartis which reached Phase II before being halted (Fig. 1).

Mixed noradrenergic profiles

Abbott have attempted to enhance the desired effects in a SSRI by introducing alpha-2 antagonistic properties into the same molecule (napitane, A-

75200) [48, 49]. Napitane proved to be effective in several animal models and initial reports suggested little or no haemodynamic effects. The compound was not better than placebo in the first efficacy study and its development has been stopped during Phase II.

Dopaminergic directed antidepressants

Single action drugs: dopamine reuptake inhibitors (DRI)

The relationship between dopamine and mood, and the mood enhancing properties of dopaminergic agents such as cocaine and amphetamine are well documented. Amphetamine itself has been used as an effective antidepressant in some patients [50], moreover, many antidepressants, e.g., desipramine, have a positive effect on the dopaminergic system after chronic but not acute treatment [51]. Thus, the dopaminergic system may be a common pathway for mood elevating drugs and activation offers hope for a rapid relief of depressed mood [52]. Inevitably, enthusiasm has been tempered because of the association of the dopaminergic system with drug abuse. However, selective dopamine reuptake inhibitors, such as buprorion, have proven to be effetive antidepressants [53]. Of the most recent attempts to develop drugs in this area, amineptine, a tricyclic derivative which is both a DRI and dopamine release enhancer, has received the most attention (Fig. 1) [54, 55]. This drug is available in a limited number of markets, and has been associated with a number of side-effects, including severe acne, and a potential for abuse. Dopaminergic drugs have suffered from an increased potential for the development of seizures, and at least one failed as an antidepressant and was retargeted towards Parkinson's disease.

Single action drugs: dopamine agonists

Although specific dopamine receptor subtypes are well known, selective dopamine agonists aimed at the psychiatry market are limited. To date only MR 708, a dopamine D2 agonist, and roxindole a dopamine D2/D3 agonist, have shown any potential for antidepressant activity [56, 57]. Compounds with similar properties are also under investigation for disorders such as Parkinson's disease.

Multiple monoaminergic action drugs

Dual action drugs: serotonin and noradrenaline

The major criticism of the single action drugs such as SSRIs is that of efficacy in severely depressed patients. Given that the tricyclics typically exert effects on several monoaminergic systems, and that combination therapies offer enhanced efficacy [15–20], it is not surprising that dual action drugs have received increasing attention. Indeed, it is now fashionable to indicate activity at more than one site, however weak.

The ability of the chemists to produce compounds with inhibitory actions at more than one uptake site initially led to the development of venlafaxine [27], a compound which acts both as a SSRI and a NRI. A series of dual action uptake inhibitors have subsequently followed, with varying ratios of 5HT to noradrenaline uptake. For example, duloxetine and milnacipran are dual 5HT and NA reuptake inhibitors [28, 58] (Fig. 1), however, milnacipran has a preferential activity at the NA transporter, and duloxetine at the 5HT transporter. These compounds may suffer from the combined side-effect profile of the serotonergic and noradrenergic selective drugs, but their dual action also seems to convey an advantage in efficacy over the SSRIs in clinical trials. Moreover, the onset of action for venlafaxine may be quicker than that observed with some of the currently available antidepressants [59]. However, a meaningful effect for the patient is still measured in weeks rather than days. Duloxetine is no longer in development as an antidepressant but both venlafaxine and milnacipran are now marketed.

The improved efficacy of dual action drugs is further supported by comparative studies with mirtazapine, a noradrenergic and specific serotonin antagonist (NaSSA), and fluoxetine [60]. Mirtazapine showed greater efficacy and a faster onset of action than fluoxetine [29]. Mirtazapine's dual activity results from a combination of selective CNS presynaptic alpha-2 blockade leading to enhanced noradrenaline and serotonin release; however, the resultant increased serotonin release is targeted towards 5HT1A because mirtazapine is also an effective 5HT2A, 5HT2C and 5HT3 antagonist. In addition to achieving the same improvements in efficacy reported with dual action reuptake inhibitors, this novel combination allows a significant reduction in side-effects to that observed with the pharmacologically less specific SSRIs or dual action reuptake inhibitors.

The mechanism of action of mirtazapine, i.e., selective 5HT1A stimulation and alpha-2 blockade, has also been pursued by Pfizer in their compound sunepitron [61]. Sunepitron is a 5HT1A agonist and alpha-2 antagonist; currently in Phase II clinical trials, sunepitron would be expected to have similar efficacy and offer similar advantages to mirtazapine over the reuptake inhibitors and TCA's.

Dual action drugs: serotonin and dopamine

Dual action drugs at both the dopaminergic and serotonergic systems are less in evidence than combination 5HT/NA drugs. An exception is roxindole, a dopamine D2/D3 agonist with additional 5HT1A agonist and SSRI properties [56, 57]. The combination of direct action at both dopamine and serotonin sites might be expected to lead to a fast onset of action, and is suggested by early trials [57]. A bupropion/SSRI combination has also been shown to increase efficacy over either drug alone, reinforcing the principle that multiple action drugs may be more efficacious than single acting drugs [62].

Triple action drugs: serotonin, noradrenaline and dopamine

If two actions are better than one, then three actions may be better than two. Tempering the effects of dopaminergic stimulation by concomitant actions at noradrenergic and serotonergic sites has been followed by several groups. The knowledge gained from the distinct reuptake sites has been used to adapt compounds with selectivity at one or two sites to form drugs acting as serotonin, noradrenaline and dopamine reuptake inhibitors (e.g., Org 32782, NS 2389) [63]. None of these triple action drugs has yet passed Phase I clinic, however, results will be eagerly awaited.

Potential non-monoaminergic antidepressant drugs

It is a sobering point that after 40 years of research into biogenic amines, we have not yet produced drugs of greater efficacy or with a substantially quicker onset of action than the TCAs. The reuptake revolution has effectively, for the research, come and gone. They have delivered tolerable and effective drugs for many patients, but still leave too many untouched. Modern dual action drugs have achieved the efficacy of the old tricyclics with much improved tolerability and lack of side-effects. Directly acting agonists, and triple action drugs have yet to prove their value in the clinic.

The anticipated improvements in monoamine based therapies discussed above are likely to yield beneficial but incremental changes in therapeutic effect, rather than any fundamental shift in drug performance. Fundamental changes in therapy are only likely to occur with a radical shift in emphasis away from monoamines. Many of the current candidates for new and potentially third generation antidepressants are indicated in Figure 2. In contrast to the monoamines, few compounds are close to the clinic use and the rationale is based on research into mechanisms and potential usefulness.

Modulation of the HPA axis: corticotropin releasing factor (CRF)

Major disturbances in the HPA axis are a common feature of depression and it has been postulated that a dysfunction in the HPA axis could be fundamental in the etiology of depression [64, 65]. The normal feedback system for control of cortisol is frequently impaired: levels of cortisol are high and fail to show normal suppression in response to dexamethasone adminstration and there is a blunted secretion of ACTH in response to CRF. There is some evidence that the more severely depressed patients exhibit greater disturbances. Moreover, correction of the hypercortisolaemia parallels recovery from depression. Reintroducing feedback into the HPA axis and restoring normal function may therefore quickly alleviate depression. Most work in this field has concentrated on the development of CRF antagonists. CRF, released in response to stress, stimulates the release of ACTH and thereby induces the release of cortisol. Accordingly, CRF antagonists should reduce cortisol levels and allow the apparently underreactive and downregulated glucocorticoid receptors (GR) to recover normal function. Chemists have had to overcome the major obstacles of synthesising stable non-peptide antagonists, which act both centrally and at the level of the pituitary. Animal studies have suggested some positive effects of the antagonists, in particular, anxiolytic effects have been noted

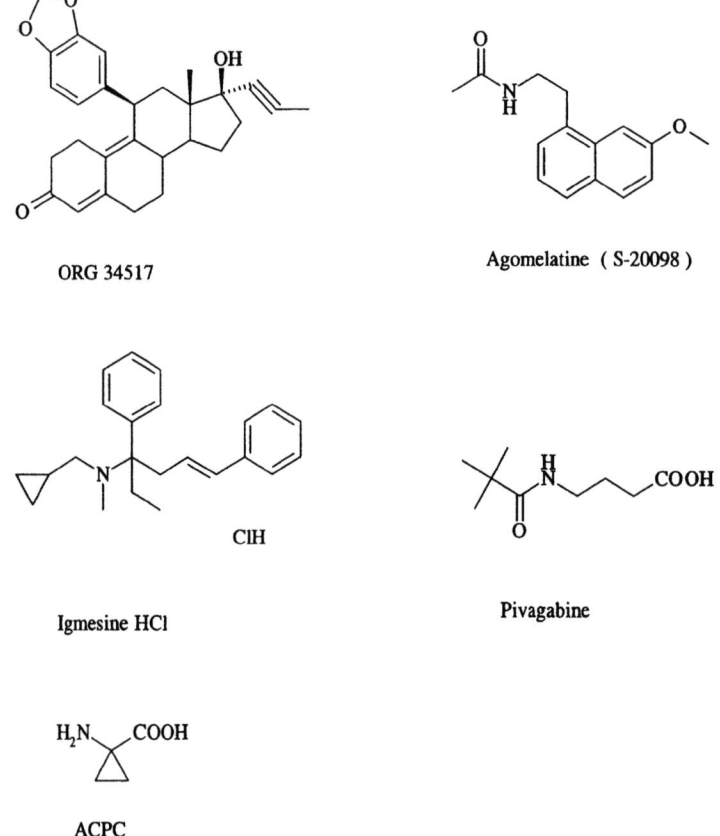

Figure 2. New putative antidepressants not based upon monoaminergic mechanisms of action.

[66]. Although almost all the major companies have programmes in this area, few compounds have made it to the clinic and no clinical trials have been fully reported. The collaboration between Neurocrine Biosciences and Jansen pharmaceuticals involving NBI 30775/R-121919 appears to be the most developed having completed Phase I and due to enter Phase II. In Italy, a CRF modulator has been launched by Angelini [67], pivagabine, (Fig. 2), but little is known about its exact mechanism of action.

Modulation of the HPA axis: glucocorticoid receptor antagonists

Steroid synthesis inhibitors can rapidly alleviate depression in patients suffering from Cushings disease, where high steroid levels and depression are commonly associated [65]. This form of therapy effectively removes the overstimulation of the GR receptors by dramatically lowering cortisol levels. However, the use of steroid synthesis inhibitors, such as ketoconazole, is not ideal for clinical use (but see [65]). A GR antagonist could be used to temporarily block GRs thereby reducing cortisol-mediated effects at target cells in the brain. Potentially, GR blockade could lead to up-regulation or increased sensitisation of the GR receptors, thus, when the GR antagonist is removed the receptors may then be sufficiently responsive to reinstate normal feedback. To date only a few selective GR antagonists have been described, only one is currently at a clinical phase, Org 34517 (Fig. 2). However, an earlier open study using RU 486, which in addition to its antiprogesterone properties is also an effective GR antagonist, demonstrated some encouraging antidepressant-like effects [65].

Modulation of the HPA axis: melatonin

Melatonin has enjoyed much attention of late for its purported effects on jet lag and its involvement in sleep regulation. Sleep disturbances and changes in circadian function were the earliest characteristics of depression to be described and investigated (reviewed in [68]). Links between melatonin and depression are controversial and in general the use of melatonin does not seem to have a large impact on the field of depression, but at least one derivative is under investigation as an antidepressant (agromelatine) [69]. The close links between melatonin, circadian disturbances, CRF and the HPA system do however suggest some scope for beneficial effects [70].

Neurokinin 1

The discovery that a Substance P antagonist (MK 869) could be an effective antidepressant is the first significant move away from the dominance of monoamine based drug therapies [71]. Substance P may play a coordinating role in the interaction between monaminergic pathways involved in modulating stress responses (e.g. [72]). Unfortunately the exact structure of MK 869 has not yet been released, although it is undoubtedly disclosed along with other members of the series in a recent patent. However, the excitement generated by this compound has to be tempered by the paucity of clinical data. The initial study showed MK 869 to be as effective as an SSRI, and to possess a similar time-course for onset of action. Therefore, whether this compound heralds the long awaited breakthrough in efficacy and speed of onset has yet to be determined. Several other NK1 antagonists are in development and optimism remains high as to the potential for novel antidepressants from this class, or related classes of NK receptors.

Modulation of glutamate: glycine antagonists

The structural considerations of developing molecules with discrete effects on monoaminergic function are relatively well understood, and have dominated the design of antidepressant drugs. No rational basis in terms of drug efficacy seems to favour one biogenic amine over another, and the greater efficacy of combination treatments or dual action drugs suggests an additive action on a separate common system.

It is commonly believed that the slow onset of action of antidepressant effects are dependent on adaptive changes occurring in the brain, i.e. that changes in organisation and neural plasticity must occur before a significant antidepressant action can be observed. The neurotransmitter system primarily associated with neural plasticity in the brain is the glutamatergic system, and especially the NMDA system [73]. NMDA interacts with all the classical monoamine transmitters, and NMDA mediated effects are altered by acute and chronic stress [74–76]. Moreover, classical antidepressants alter NMDA subtype composition in the CNS [77]. Intriguingly, these changes occur only after chronic antidepressant treatment, and are also apparent after electroconvulsive shock [78, 79].

Both direct and indirect NMDA antagonists have demonstrated positive effects in animal models [80, 81], however, the relationship of NMDA blockade to the induction of hallucinations probably renders them impractical for drug design. Instead allosteric modulators of NMDA function are the favoured target. Particular attention has been paid to ligands acting at the glycine site on the NMDA receptor, occupation of which is necessary for normal NMDA function [82]. Glycine antagonists can produce antidepressant-like changes in NMDA function [83], manifest as a change in the ability of glycine to bind to the NMDA receptor thereby altering its activation. Antidepressant-like effects have been reported for several glycine antagonists, such as ACPC [84] (Fig. 2), currently the only glycine antagonist close to clinical use (Phase I). Undoubtedly, other glycine site ligands will follow for affective disorders, meanwhile glycine uptake inhibitors are also being investigated for psychosis. Interestingly, glycine modulates the effects of NK1 receptors on the NMDA system suggesting some link between the actions of these drugs [85].

The NMDA receptor is a complex ligand-gated ion channel possessing a large number of allosteric modulatory sites dependent on the subunit composition. It offers a large target for medicinal chemists in its own right. Of more fundamental importance is that for the first time changes in gene expression and RNA levels in the major transmitter system in the CNS can be linked to antidepressant activity. It is elucidating the mechanism by which these types of changes occur which will allow molecular biology to identify the novel targets for effective drug therapy.

Sigma ligands

Sigma ligands have been linked to psychiatric disorders for some time. Initially, sigma ligands were linked to opiates and psychosis [86]; unfortunately, clinical efficacy was not forthcoming. Moreover, structure activity relationships in sigma ligands have proven difficult, and much uncertainty surrounds the status of the receptor, even following its recent cloning [87] (indeed some interactions may even be *via* enzyme modulation at an enzyme site rather than a receptor [88]). One compound, igmesine (Fig. 2), has both opiate antagonist and sigma blocking properties. Preclinical research indicated positive antidepressant-like effects in some models [89], and a recent report suggests this may translate to some efficacy in depressed patients (paper presented by Pande et al., CINP, Glasgow, 1998).

Non-drug treatments

Despite the variety of drugs under development, the aim of substantial increases in efficacy and speed of onset still appears distant. However, therapies are available which indicate that these are demonstrably attainable goals. ECT remains the most effective antidepressant treatment for severely depressed patients with a poor response to drug therapy; it is also substantially quicker in onset [90]. In addition, sleep deprivation leads to a rapid, though transient, alleviation of depressed mood [91]. Identifying the mechanism of action of either therapy would lead to substantial improvements in drug therapy.

The time required for antidepressants to become effective suggests that simple changes in monoamines are insufficient to counter depressive mood. Adaptive changes must occur over the course of weeks *via* changes in second messenger systems within the cell, or *via* indirect effects on other neurotransmitter systems such as the glutamate system. Whatever the route, all these changes require changes in the transcription of genes. Research into identifying relevant changes in gene expression and transcription factors is underway. Identifying genes in families with depression has produced tantalising if, as yet, inconclusive results. The convergence of genetic linkage studies with molecular biological techniques such as differential display suggests a promising future for research and for radically new targets for drugs.

A new pharmacology for the third generation

The development of antidepressant drugs over the last 40 years revealed important insights into the pharmacology of depression. However, information on the pharmacology of the drugs has not necessarily resulted in a clear understanding as to the causes of depression. For example, depression is not an automatic consequence of serotonin depletion, and normal subjects do not develop depression after serotonin (or indeed catecholamine) depletion [92]. As a tool, serotonin depletion has proven interesting, but merely reinforced fashionable assumptions as to the core causes of depression without actually producing more effective antidepressants. Severe depression is also observed after the administration of cholinergic agonists [93], however, the potential involvement of the cholinergic system in depression has received relatively little attention. Perhaps of greater interest is the induction of depression in patients undergoing cytokine therapy during cancer treatment. This effect is such a predictable consequence of treatment that the AMA has issued official guidelines that patients involved in such a treatment regime should also receive concurrent antidepressant therapy. These observations suggest that the link between stress, the HPA axis and the immune system may be more profitably explored than at present (see [94, 95] for relevant reviews).

Patients may respond well to one drug and not to another. Similarly, patients exhibit a wide range of symptoms which may reflect different underlying causes. A greater understanding of the patients and the symptoms may lead to a more focussed and effective drug treatment for individual patients. Equally evident is that antidepressants are no longer just for depression. Accordingly, an understanding of the pharmacology of antidepressants, and the interactions of the distinct serotonin, noradrenergic and dopaminergic receptors encourages the use and development of drugs targeted towards symptoms associated with modulation of the neurotransmitter system irrespective of the psychiatric disorder. Similar symptoms appear in several disorders, for example, anxiety or sleep disturbances in GAD, depression and schizophrenia. Some atypical antipsychotics are potentially more effective than the older dopamine antagonists in treating problems associated with schizophrenia, these compounds are also serotonin antagonists, properties inherent in many new antidepressants. Integrating the knowledge we have on the pharmacology of the newer drugs suggests a role for utilising specific combinations of drugs to produce more favourable treatment strategies for patients exhibiting multiple symptoms.

Thus, the true legacy of the late second generation compounds may extend beyond the confines of depression. Exploiting our understanding of the mechanism of action of drugs in influencing aspects of brain and behaviour will inevitably lead to improvements in treatment strategies for psychiatric patients. The compounds reviewed here represent the first step towards a broader understanding of the pharmacology of psychiatric drugs. Meanwhile, advances in genetics, second messenger systems and gene transcription factors will drive progress in formulating the highly efficacious and rapidly acting drugs of the future.

References

1 Greenberg PE, Stiglin LE, Finkelstein SN, Berndt ER (1993a) The economic burden of depression in 1990. *J Clin Psych* 54: 403–418
2 Greenberg PE, Stiglin LE, Finkelstein SN, Berndt ER (1993b) Depression: a neglected major illness. *J Clin Psych* 54: 419–423
3 Angst J (1992) Epidemiology of depression. *Psychopharmacology* 106: S71–S74
4 Lepine J-P, Gastpar M, Mendlewicz J, Tylee A, on behalf of the DEPRES Steering Committee (1997) Depression in the community: the first panEuropean study DEPRES (Depression Research in European Society). *Internat Clin Psychopharma* 12: 19–29
5 Pinder RM (1988) The benefits and risks of antidepressant drugs. *Human Psychopharma* 3: 73–86
6 Pinder RM (1996) Why more antidepressants? *Pharma News* 3 (4): 11–14
7 Pinder RM, Wieringa JH (1993) Third-generation antidepressants. *Med Res Rev* 13: 259–325
8 Anderson IM, Tomenson BM (1994) The efficacy of selective serotonin reuptake inhibitors in depression: a meta-analysis of studies against tricyclic antidepressants. *J Psychopharma* 8: 238–249
9 Schou M (1997) Forty years of lithium treatment. *Arch Gen Psych* 54: 9–13
10 Post RM, Ketter TA, Denicoff K, Pazzaglia PJ, Leverich GS, Marangeii LB, Callahan AM, George MS, Frye MA (1996) The place of anticonvulsant therapy in bipolar illness. *Psychopharma* 128: 115–129
11 Broekkamp CLE, Leysen D, Peeters PWMM, Pinder RM (1995) Prospects for improved antidepressants. *J Med Chem* 38: 4615–4633
12 Skolnick P (ed) (1997) *Antidepressants. New pharmacological strategies*. Humana Press, Totowa, New Jersey
13 Blier P, De Montigny C (1994) Current advances and trends in the treatment of depression. *Trends Pharma Sci* 15: 220–226
14 Thase ME, Rush AJ, Kasper S, Nemeroff CB (1995) Tricyclics and newer antidepressant medications: treatment options for treatment-resistant depressions. *Depression* 3: 152–168
15 Nemeroff CB (1997) Augmentation strategies in patients with refractory depression. *Depression Anxiety* 4: 169–181
16 Artigas F, Romero L, De Montigny C, Blier P (1996) Acceleration of the effect of selected antidepressant drugs in major depression by 5-HT1A antagonists. *Trends Neuro Sci* 19: 378–383
17 Nelson JC, Mazure CM, Bowers MB, Jatiow PL (1991) A preliminary open study of the combination of fluoxetine and desipramine for rapid treatment of major depression. *Arch Gen Psych* 48: 303–307

18 Dam J, Ryde O, Svejso J, Lauge N, Lauritzen B, Bech P (1998) Morning fluoxetine plus evening mianserin vs. morning fluoxetine plus evening placebo in the acute treatment of major depression. *Pharmacopsychiatry* 31: 1–7

19 Maes M, Libbrecht I, van Hunsel F, Campens D, Meltzer HY (1999) Pindolol and mianserin augment the antidepressant efficacy of fluoxetine in hospitalised major depressed patients even in those with treatment resistance. *J Clin Psychopharma* 19: 177–182

20 Cappiello A, Oren D, Anand A, Berman R, Charney D (1998) Yohimbine plus fluoxetine combination for rapid treatment of depression. Abstract 36th Meeting of American College of Neuropsychopharmacology, Hawaii, 250

21 Pinder RM (1997) Designing a new generation of antidepressant drugs. *Acta Psychiatrica Scandinavica* 96 (Suppl. 391): 7–13

22 Sambunaris A, Keppel Hesselink J, Pinder R, Panagides J, Stahl SM (1997) Development of new antidepressants. *J Clin Psych* 58 (Suppl. 6): 40–53

23 Nutt DJ (ed) (1977) Reboxetine: a selective noradrenaline reuptake inhibitor. Additional benefits to the depressed patient. *J Psychopharma* 11 (Suppl): S3–S47

24 Romero L, Bel N, Casanovas JM, Artigas F (1996) Two actions are better than one: avoiding self-inhibition of serotonergic neurons enhances the effects of serotonin uptake inhibitors. *Internat Clin Psychopharma* 11 (Suppl 4): 1–8

25 Shader RL, Fogelman SM, Greenblatt DJ (1997) Newer antidepressants: hypotheses and evidence. *J Clin Psychopharma* 17: 1–3

26 Stahl SM (1997) Are two antidepressant mechanisms better than one? *J Clin Psych* 58: 339–340

27 Clerc GE, Ruimy P, Verdeau-Pailles J on behalf of the Venlafaxine French Inpatient Study Group (1994) A double-blind comparison of venlafaxine and fluoxetine in patients hospitalised for major depression and melancholia. *Internat Clin Psychopharma* 9: 139–143

28 Lopez-Ibor J, Guelfi JD, Pletan Y, Tournoux A, Prost JF (1996) Milnacipran and selective serotonin reuptake inhibitors in major depression. *Internat Clin Psychopharma* 11 (Suppl 4): 41–46

29 Wheatley D, Van Moffaert M, Timmerman L, Kremer CE for the Mirtazapine-Fluoxetine Study Group (1998) Mirtazapine: efficacy and tolerability in comparison with fluoxetine in patients with major depression. *J Clin Psych* 59: 306–312

30 Bianchi G, Caccia S, Della Vedova F, Garattini S (1988) The alpha-2-adrenoceptor antagonist activity of ipsapirone and gepirone is mediated by their common metabolite 1-(2-pyrimidinyl)-piperazine (PmP). *Eur J Pharmacol* 151: 365–371

31 Feiger AD (1996) A double-blind comparison of gepirone extended release, imipramine, and placebo in the treatment of outpatient major depression. *Psychopharmacol Bull* 32: 659–665

32 McGrath PJ, Stewart JW, Quitkin FM, Wager S, Jenkins SW, Archibald DG, Stringfellow JC, Robinson DS (1994) Gepirone treatment of atypical depression: preliminary evidence of serotonergic involvement. *J Clin Psychopharmacol* 14: 347–352

33 Cryan JF, Redmond AM, Kelly JP, Leonard BE (1997) The effects of the 5-HT1A agonist flesinoxan, in three paradigms for assessing antidepressant potential in the rat. *Eur Neuropsychopharmacol* 7: 109–114

34 Dugast L (1998) Is the potent 5-HT1A receptor agonist, alnespirone (S-20499), affecting dopaminergic systems in the rat brain? *Eur J Pharmacol* 350: 171–180

35 Costall B, Domeney AM, Farre AJ, Kelly ME, Martinez L, Naylor RJ (1992) Profile of action of a novel ligand E-4424 to inhibit aversive behavior in the mouse, rat and marmoset. *J Pharmacol Exp Ther* 262: 90–98

36 Johansson L, Sohn D, Thorberg SO, Jackson DM, Kelder D, Larsson LG, Renyi L, Ross SB, Wallsten C, Eriksson H et al (1997) The pharmacological characterization of a novel selective 5-hydroxytryptamine1A receptor antagonist, NAD-299. *J Pharmacol Exp Ther* 283: 216–225

37 Berendsen HH, Broekkamp CL (1994) Comparison of stimulus properties of fluoxetine and 5-HT receptor agonists in a conditioned taste aversion procedure. *Eur J Pharmacol* 253: 83–89

38 Martin JR, Bos M, Jenck F, Moreau J, Mutel V, Sleight AJ, Wichmann J, Andrews JS, Berendsen HH, Broekkamp CL et al (1998) 5-HT2C receptor agonists: pharmacological characteristics and therapeutic potential. *J Pharmacol Exp Ther* 286: 913–924

39 Jenck F, Moreau JL, Berendsen HH, Boes M, Broekkamp CL, Martin JR, Wichmann J, Van Delft AM (1998) Antiaversive effects of 5HT2C receptor agonists and fluoxetine in a model of panic-like anxiety in rats. *Eur Neuropsychopharmacol* 8: 161–168

40 Mellow AM, Lawler BA, Sunderland T, Mueller EA, Molchan SE, Murphy DL (1990) Effects of daily oral m-chlorophenylpiperazine in elderly depressed patients; initial experience with a serotonin agonist. *Biol Psychiatry* 28: 588–594

41 Kennett GA, Wood MD, Bright F, Trail B, Riley G, Holland V, Avenell KY, Stean T, Upton N, Bromidge S, Forbes IT, Brown AM, Middlemis DN, Blackburn TP (1997) SB 242084, a selective and brain penetrant 5-HT2C receptor antagonist. *Neuropharmacology* 36: 609–620

42 Davis R, Whittington R, Bryson HM (1997) Nefazodone. A review of its pharmacology and clinical efficacy in the management of major depression. *Drugs* 53. 608–636

43 Bakish D, Hooper CL, Thornton MD, Wiens A, Miller CA, Thibaudeau CA (1997) Fast onset: an open study of the treatment of major depressive disorder with nefazodone and pindolol combination therapy. *Int Clin Psychopharmacol* 12: 91–97

44 D'Aquila P, Monleon S, Borsini F, Brain P, Willner P (1997) Anti-anhedonic actions of the novel serotonergic aagent flibanserin, a potential rapidly-acting antidepressant. *Eur J Pharmacol* 340: 121–132

45 Borsini F, Cesana R, Kelly J, Leonard BE, McNamara M, Richards J, Seiden L (1997) BIMT 17: a putative antidepressant with a fast onset of action? *Psychopharmacology* 134: 378–386

46 Montgomery SA (1997) Reboxetine: additional benefits to the depressed patient. *J Psychopharmacol* 11 Suppl 4: S9–15

47 Trendelenburg AU, Wahl CA, Starke K (1996) Antagonists that differentiate between alpha(2A)- and alpha(2D)-adrenoceptors. *Naunyn-Schmiedebergs Arch Pharmacol* 353: 245–249

48 Firestone JA, Gerhardt GA, DeBernardis JF, McKelvy JF, Browning MD (1993) Actions of A-75200, a novel catecholamine uptake inhibitor, on norepinephrine uptake and release from bovine adrenal chromaffin cells. *J Pharmacol Exp Ther* 264: 1206–1210

49 Hancock AA, Buckner SA, Giardina WJ, Brune ME, Lee JY, Morse PA, Oheim KW, Stansic DS, Warner RB, Kerkman DJ et al (1995) Preclinical pharmacological actions of (+/–)-(1′R*,3R*)-3-phenyl-1-[1′2′,3′,4′-tetrahydro-5′,6′-methylene-dioxy-1′-naphthalenyl) methyl] pyrrolidine methanesulfonate (ABT-200), a potential antidepressant agent that antagonizes alpha-2 adrenergic receptors and inhibits the neuronal uptake of norepinephrine. *J Pharmacol Exp Ther* 272: 1160–1169

50 Satel SL, Nelson JC (1989) Stimulants in the treatment of depression: a critical overview. *J Clin Psychiatry* 50: 241–249

51 Fibiger HC, Phillips AG (1981) Increased intracranial selfstimulation in rats after long-term administration of desipramine. *Science* 214: 683–685

52 Willner P (1997) The mesolimbic dopamine system as a target for rapid antidepressant action. *Int Clin Psychopharmacol* 12 Suppl 3: S7–S14

53 Weisler RH, Johnston JA, Lineberry CG, Samara B, Branconnier RJ, Billow AA (1994) Comparison of bupropion and trazadone for the treatment of major depression. *J Clin Psychopharmacol* 14: 170–179

54 Dalery J, Rochat C, Peyron E, Bernard G (1997) The efficacy and acceptability of amineptine vs. fluoxetine in major depression. *Int Clin Psychopharmacol* 12 Suppl 3: S35–S38

55 Garattini S (1997) Pharmacology of amineptine, an antidepressant agent acting on the dopaminergic system: a review. *Int Clin Psychopharmacol* 12 Suppl 3: S15–S19

56 Maj J, Kolodziejczyk K, Rogoz Z, Skuza G (1997) Roxindole, a dopamine autoreceptor agonist with a potential antidepressant activity. II. Effects on the 5-hydroxytryptamine system. *Pharmacopsychiatry* 30: 55–61

57 Grunder G, Wetzel H, Hammes E, Benkert O (1993) Roxindole, a dopamine autoreceptor agonist, in the treatment of major depression. *Psychopharmacology* 111: 123–126

58 Kihara T, Ikeda M (1995) Effects of duloxetine, a new serotonin and norepinephrine inhibitor, on extracellular monoamine levels in rat frontal cortex. *J Pharmacol Exp Ther* 272: 177–183

59 Derivan A, Entsuah A, Kikta D (1995) Venlafaxine: measuring the onset of antidepressant action. *Psychopharmacology* 131: 349–447

60 Stimmel GL, Dopheide JA, Stahl S (1997) Mirtazapine: an antidepressant with noradrenergic and specific serotonergic effects. *Pharmacoptherapy* 17: 10–21
61 Silvestre J, Graul A, Castaner J (1998) Sunepitron hydrochloride. *Drugs of the Future* 23: 161–165
62 Bodkin JA, Lasser RA, Wines JD, Gardner DM, Baldessarini RJ (1997) Combining serotonin reuptake inhibitors and bupropion in partial responders to antidepressant monotherapy. *J Clin Psychiatry* 58: 137–145
63 Carlier P, Lo M, Lo P, Richelson E, Atsumi M, Reynolds I, Sharma T (1998) Synthesis of a potent wide-spectrum serotonin-, norepinephrine-, dopamine-reuptake inhibitor (SNDRI) and a species-selective dopamine reuptake inhibitor based on the gamma-amino alcohol functional group. *Bioorganic & Medicinal Chemistry Letters* 8: 487–492
64 Holsboer F, Barden N (1996) Antidepressants and the hypothalamic pituitary adrenocortical regulation. *Endocr Rev* 17: 187–205
65 Murphy BE (1997) Antiglucocorticoid therapies in major depression: a review. *Psychoneuroendocrinology* 22 Suppl 1: S125–132
66 Behan DP, Grigoriadis DE, Lovenberg T, Chalmers D, Heinrichs S, Liaw C, De Souza EB (1996) Neurobiology of corticotropin releasing factor (CRF) receptors and CRF binding protein: implications for the treatment of CNS disorders. *Mol Psychiatry* 1: 265–277
67 Esposito G, Luparini MR (1997) Pivagabine: a novel psychoactive drug. *Arzneimittelforschung* 47: 1306–1309
68 Gold PW, Goodwin FK, Chrousos (1988) Clinical and biochemical manifestations of depression. Relation to the neurobiology of stress. *New Eng J Med* 319: 413–420
69 Martinet L, Guardiola-Lemaitre B, Mocaer E (1996) Entrainment of circadian rhythms by S-20098, a melatonin agonist, is dose and plasma concentration dependent. *Pharmacol Biochem Behav* 54: 713–718
70 Kellner M, Yassouridis A, Manz B, Steiger A, Holsbooe F, Wiedemann K (1997) Corticotropin-releasing hormone inhibits melatonin secretion in healthy volunteers – a potential link to low-melatonin syndrome in depression? *Neuroendocrinology* 65: 284–290
71 Kramer S, Cutler N, Feighner J, Shrivastava R, Carman J, Sramek JJ, Reines SA, Liu G, Snavely D, Wyatt-Knowles E et al (1998) Distinct mechanism for antidepressant activity by blockade of central substance P receptors. *Science* 281: 1640–1645
72 Horger BA, Roth RH (1996) The role of mesoprefrontal dopamine neurons in stress. *Crit Rev Neurobiol* 10: 395–418
73 McBain CJ, Mayer ML (1994) N-methyl-D-aspartic acid receptor structure and function. *Physiol Rev* 74: 723–760
74 Nowak G, Redmond A, McNamara M, Paul IA (1995) Swim stress increases the potency of glycine at the N-methyl-D-aspartate receptor complex. *J Neurochem* 64: 925–927
75 Kim JJ, Foy MR, Thompson RF (1996) Behavioral stress modifies hippocampal plasticity through N-methyl-D-aspartate receptor activation. *Proc Natl Acad Sci USA* 93: 4750–4753
76 Loscher W, Honack D (1992) The behavioural effects of MK-801 in rats: involvement of dopaminergic, serotonergic and noradrenergic systems. *Eur J Pharmacol* 215: 199–208
77 Paul IA, Nowak G, Layer RT, Popik P, Skolnick P (1994) Adaptation of the N-Methyl-D-Aspartate receptor complex following chronic antidepressant treatments. *J Pharmacol Exp Ther* 269: 95–102
78 Paul IA, Layer RT, Skolnick P, Nowak G (1993) Adaption of the N-Methyl-D-Aspartate receptor complex in rat front cortex following chronic treatment with electroconvulsive shock or imipramine. *Eur J Pharmacol* 247: 305–312
79 Nowak G, Li Y, Paul IA (1996) Adaptation of cortical but not hippocampal NMDA receptors after chronic citalopram treatment. *Eur J Pharmacol* 295: 75–85
80 Maj J, Rogoz Z, Skuza G, Sowinska H (1992) The effect of CGP 37849 and CGP 39551, competitive NMDA receptor antagonist, in the forced swimming test. *Pol J Pharmacol Pharm* 44: 337–346
81 Maj J, Rogoz Z, Skuza G, Sowinska H (1992) Effects of MK-801 and antidepressant drugs in the forced swimming test in rats. *Eur Neuropsychopharmacol* 2: 37–41
82 Thomson AM (1990) Glycine is a coagonist at the NMDA receptor/channel complex. *Prog Neurobiol* 35: 53–74
83 Fossom LH, Basile AS, Skolnick P (1995) Sustained exposure to 1-aminocyclopropanecarboxylic acid, a glycine partial agonist, alters N-methyl-D-aspartate receptor function and subunit composition. *Mol Pharmacol* 48: 981–987

84 Papp M, Moryl E (1996) Antidepressant-like effects of 1-aminocyclopropanecarboxylic acid and D-cycloserine in an animal model of depression. *Eur J Pharmacol* 316: 145–151
85 Heppenstall PA, Fleetwood-Walker SM (1997) The glycine site of the NMDA receptor contributes to the neurokinin 1 receptor agonist facilitation of NMDA receptor agonist-evoked activity in rat dorsal horn neurons. *Brain Res* 744: 235–245
86 Debonnel G (1993) Current hypotheses on sigma receptors and their physiological role: possible implications in psychiatry. *J Psychiatry Neurosci* 18: 157–172
87 Seth P, Fei YJ, Li HW, Huang W, Leibach FH, Ganapathy V (1998) Cloning and functional characterisation of a sigma receptor from rat brain. *J Neurochem* 70: 922–931
88 Moebius FF, Bermoser K, Reiter RJ, Hanner M, Glossman H (1996) Yeast sterol C8-C7 isomerase: identification and characterization of a high-affinity binding site for enzyme inhibitors. *Biochemistry* 35: 16871–16878
89 Matsuno K, Kobayashi T, Tanaka MK, Mita S (1996) Sigma 1 receptor subtype is involved in the relief of behavioral despair in the mouse forced swimming tes. *Eur J Pharmacol* 312: 267–271
90 Rich CL, Spiker DG, Jewell SW, Neil JF, Black NA (1984) The efficiency of ECT: I. Response in depressive episodes. *Psychiatric Res* 11: 167–176
91 Wu JC, Bunney WE Jr (1990) The biological basis of an antidepressant response to sleep deprivation and relapse: review and hypothesis. *Am J Psychiatry* 147: 14–21
92 Delgado PL, Moreno FA, Potter R, Gelenberg AJ (1998) Norepinephrine and serotonin in antidepressant action: evidence from neurotransmitter depletion studies. In: M Briley, S Montgomery (eds): *Antidepressant therapy at the dawn of the third millenium*. Martin Dunitz Ltd, UK, 141–161
93 Davis KL, Hollister E, Overall J, Johnson A, Train K (1976) Physostigmine: effects on cognition and affect in normal subjects. *Psychopharmacologia* 51: 23–27
94 Malek-Ahmadi P (1996) Neuropsychiatric aspects of cytokines research: an overview. *Neurosci Biobehav Rev* 20: 359–365
95 Weigers G, Reul JMHM (1998) Induction of cytokine receptors by glucocorticoids: functional and pathological significance. *TiPS* 19: 317–321

Index

acetylcholine 96
ACPC 138
ACTH 84, 94
ACTH/cortisol 96
acute phase protein 112
adrenal gland volume 89
α_2-adrenoceptor 97
α_2-adrenoceptor antagonist 96, 97
β-adrenoceptor 65, 66, 72, 97
AF 2968 131
aggression, footshock-induced 66
agromelatine 137
alcohol abuse/dependence (AAD) 6
allergy 112
alnespirone 130
alpha 2 antagonist 31
alpha 2 blockade 134
alpha 2 blocker mianserin 132
alpha 2 ligand 132
amineptine 133
aminoglutethimide 90
amitriptyline 23
amphetamine 133
antagonist, calcium channel 68
anterior pituitary gland 95
antidepressant, mechanism of action 49
antidepressant, monoamine oxidase inhibitor (MAOI) 63, 64
antidepressant, second generation 124
antidepressant, serotonin reuptake inhibitor 63
antidepressant, speed of onset 47
antidepressant, third generation 125, 126
antidepressant, tricyclic (TCA) 5, 31, 63, 64, 66, 69, 71, 98, 124
antisocial personality disorder (ASP) 6
apomorphine 96, 99
arginine vasopressin (AVP) 83
astrocyte 110
augmentation strategy 125
auto-immune disease 112

BAY K 8644 68
B-cell 113
Beck Depression Inventory 5

beta-endorphin 86
bipolar 41
body temperature 96
brain derived neurotrophic factor (BDNF) 33
bulbectomy, olfactory 66
buproprion 133
buspirone 100, 130

calcium 68, 69, 71–73
calcium channel, antagonist 48
calmodulin 69, 72, 73
cancer 14
carbamazepine 124
catecholamine 96
cellular immunity 109
CGP-36953 67
channel, L-type calcium 68
cholesterol 16
CIDI-Auto 5
circumventricular organs 111
citalopram 100
clomipramine 100
clonidine 96, 97
Cluster A 7
Cluster B 7
Cluster C 7
co-medication 27
comorbidity with personality disorders 3
comorbidity, axis I – psychiatric disorders 3
comorbidity, axis II – personality disorders 7
comorbidity, axis III – somatic disorders 9
comorbidity, psychiatric 3
comorbidity, somatic 3
complement protein 112
cortex, frontal 67
corticosterone 102
corticotropin (ACTH) 84, 97
corticotropin releasing factor (CRF) 109, 135
corticotropin releasing factor (CRF), antagonist 136
corticotropin releasing hormone (CRH) 83

cortisol 15, 84, 97, 109
cortisol, urinary free 86
CRH test 87
CSF 5-HIAA 101
cyclase, guanylyl 69, 72, 73
cyclic guanosine monophosphate (cGMP) 69, 70
cytochrome P450 enzyme system 27
cytokine, proinflammatory 109
cytokine receptor, soluble 112
cytokine therapy 140

D_2 receptor 99
"delayed onset of action", hypothesis 21
depression, macrophage theory of 109
depression, major 6
depression, pathogenesis 28
depression, refractory 22
depressive disorder, major 31
desipramine 96
dexamethasone suppression test 85
dexamethasone 15
dexamethasone/CRH test 87
diabetes 14
diagnostic entity 27
5,7-dichlorokynurenic acid (DCKA) 67
dihydropyridine 68, 69
dizocilpine 67
dopamine agonist 133
dopamine reuptake inhibitor (DRI) 133
double-blind drug trial 23
drug responder 22
dual action drug 133
duloxetine 134

estrogen 34
ethnic variation 27
excess mortality 13
expectancy of life 13

fenfluramine 100
flesinoxan 100, 130, 131
fluoxetine 23, 98
fluvoxamine 99
forced swim test 66, 68, 71, 72
forebrain, mesolimbic 100

gene expression 32
gene product 32
gene transcription factor 141
general vulnerability 9
generalised anxiety disorder (GAD) 6
genetics 141
gepirone 130

glucocorticoid receptor 84, 114
glucocorticoid receptor antagonist 137
glutamate 65, 66
glycine 66, 67
glycine antagonist 138
growth hormone (GH) 15, 97
Growth Hormone Releasing Hormone (GHRH) 98
guanosine triphosphate (GTP) 69
guanylyl, cyclase 69, 72, 73
guideline, probability-based 28

Hamilton Anxiety Scale 5
Hamilton Depression/Melancholia Scale 5, 6
health care utilisation 9, 10
helplessness, learned 66, 68
hormonal alteration 28
hormone, sex 15
5-HT 96, 125
5-HT_{1A} 100
5-HT_{1A} agonist 130
5-HT_{1A} antagonist 130
5-HT_{1A} receptor 130
5-HT_{1B} 104
5-HT_{1B} receptor 104
5-HT_{1D} 101
5-HT_{2A} agonist 31
5-HT_{2A} antagonist 31
5-HT_{2C} 100
5-HT_{2C} agonist 131
hypercortisolaemia 113
hypertension 14
hypoparathyroidism 68
hypotension 97
hypothalamic-pituitary-adrenal (HPA) axis 15, 28, 83, 110
hypothalamic-pituitary-adrenal (HPA) axis, modulation 135
hypothalamus 95

igmesine 139
imipramine 23, 64, 71, 72
immunity, cellular 109
immunoglobulin IgM 113
improvement, early onset of 24
improvement, onset 24
improvement, time-course 22–24
impulsive personality trait 101
indoleamine hypothesis 100
insulin resistance syndrome 16
insulin-like growth factor 1 (IGF-1) 16
interleukin (IL) 110
interleukin 1 (IL-1) 109
interleukin 6 (IL-6) 109
ipsapirone 100, 130
ischaemic heart disease 14

Index

ketamine 68
ketoconazole 90

L-arginine 69, 72
L-citrulline 69
lesopitron 130
life expectancy 13
lipopolysaccharide 114
lithium 124
lofepramine 105
long-term course 11
long-term outcome 13
L-tryptophan 100
L-type calcium channel 68

macrophage activity 114
macrophage theory of depression 109
major depression (MD) 6
major depressive disorder 31
maprotiline 96
m-chlorophenylpiperazine (mCPP) 100, 101, 131
melatonin 96, 137
mesolimbic forebrain 100
meta-analysis 23
metabolic syndrome 15
metoclopramide 99
metyrapone 90
microglia 110
mifepristone (RU 486) 91
milnacipran 126, 134
mirtazapine 98, 126, 134
MK 912 132
moclobemide 23
modulation of the HPA axis 135
monoamine oxidase inhibitor (MAOI) 5, 31, 124
monoaminergic system 27
mood disorder clinic 15
mood stabiliser 124
mortality 3, 13
mortality, excess 13
multiple sclerosis 112

NAD 299 131
napitane 132
natural killer cell (NKC) activity 112
NBI 30775/R-121919 137
nefazodone 131
neuroadaptive 103
neuroendocrine challenge test 95
neuroendocrine 95
neurokinin 1 137
neurotrophic factor 33
neutrophil phagocytosis 112
N^G-monomethyl-L-arginine (L-NMMA) 69, 71

N^G-nitro-L-arginine (L-NNA) 69, 71, 72
N^G-nitro-L-arginine methyl ester (L-NAME) 69, 71
nitric oxide (NO) 69–73
nitric oxide synthase (NOS) 69–73
7-nitroindazole 70
NMDA antagonist 138
NMDA receptor 65–67, 69, 70, 139
non-monoaminergic antidepressant drug 135
non-responder patient 40
noradrenaline (NA) 47, 125
noradrenaline and serotonin specific antidepressant (NASSA) 126, 134
noradrenergic and specific serotonin antagonist (NaSSA) 134
noradrenergic reuptake inhibitor (NRI) 132
noradrenergic reuptake inhibitor (NRI), desipramine 132
NS 2389 135

obstructive pulmonary disease 14
olfactory, bulbectomy 66
Org 12962 131
Org 13011 130
Org 32782 135
Org 34517 137
oxaprotiline 23

paroxetine 71
phagocytosis, neutrophil 112
pharmacogenetic study 27
Pindolol 131
pituitary hormone 95
pituitary gland, anterior 95
pivagabine 137
placebo 23
placebo responder 22
polyamine 66
polypharmacy 9, 10
potentiation, long-term 66
probability-based guideline 28
prognosis 6
proinflammatory cytokine 109
prolactin (PRL) 99
prostaglandin E_1 (PGE_1) 113
prostaglandin E_2 (PGE_2) 113
protein, acute phase 112
protein, complement 112
psychiatric health care utilisation 4
pulmonary disease, obstructive 14

reboxetine 132
recovery 37, 40
refractory depression 22

relapse 39
remission 37
response to treatment 24
reuptake inhibitor, dopaminergic 31
reuptake inhibitor, dual serotonin 31
reuptake inhibitor, dual serotonin and norepinephrine 38
reuptake inhibitor, noradrenergic 31
reuptake inhibitor, noradrenergic selective 31
reuptake inhibitor, serotonin 31
reuptake inhibitor, serotonin selective 31, 33
rheumatoid arthritis 112, 115
roxindole 133, 134
RU 486 137

Schedules for Clinical Assessment in Neuropsychiatry (SCAN) 3
SCID 6
SCID screen questionnaire 7
SDZ NVI 085 132
second generation antidepressant 124
second messenger system 141
sedation 97
selective serotonin reuptake inhibitor (SSRI) 5, 47, 98, 124, 126, 129
serotonin 47, 88
serotonin (5-HT) 96, 125
serotonin depletion 140
serotonin reuptake inhibitor (SRI) 124
sex hormone 15
shock, electroconvulsive 68
sigma ligand 139
soluble cytokine receptor 112
Standardised Mortality Ratio (SMR) 13
stress 15
stress, chronic mild 66
substance abuse disorder 6
substance P 38
substance P antagonist (MK 869) 137
suicide 67, 101
sumatriptan 100, 101

sunepitron 134
sympathoadrenal medullary system (SAM) 83
synacthen test 89
synthase, nitric oxide (NOS) 69–73

tail suspension test 66
T-cell, activated 113
test, CRH 87
test, dexamethasone suppression 85
test, dexamethasone/CRH 87
test, forced swim 66, 68, 71, 72
test, neuroendocrine challenge 95
test, tail suspension 66
T-helper cell 113
therapeutic response 22
therapy, cytokine 140
third-generation antidepressant 125, 126
thyrotropin releasing hormone 102
T-memory cell 113
treatment refractory patient 40
treatment resistant patient 40
treatment response patient 36
tricyclic antidepressant (TCA) 98, 124
triglyceride 16
triple action drug 135
tryptophan (TRP) depletion 105
tumour necrosis factor alpha (TNF-α) 109

urinary free cortisol 86

valproate 124
venlafaxine 126, 134
vulnerability, general 9

Waist to Hip Ratio (WHR) 15

yohimbine 97

If you have any concerns about our products,
you can contact us on
ProductSafety@springernature.com

In case Publisher is established outside the EU,
the EU authorized representative is:
**Springer Nature Customer Service Center GmbH
Europaplatz 3, 69115 Heidelberg, Germany**

Printed by Libri Plureos GmbH
in Hamburg, Germany